BACTERIOLOGY RESEARCH DEVELOPMENTS

BOTULINUM TOXIN

USES, MECHANISMS OF ACTION AND COMPLICATIONS

BACTERIOLOGY RESEARCH DEVELOPMENTS

Additional books in this series can be found on Nova's website under the Series tab.

Additional e-books in this series can be found on Nova's website under the e-book tab.

NEUROSCIENCE RESEARCH PROGRESS

Additional books in this series can be found on Nova's website under the Series tab.

Additional e-books in this series can be found on Nova's website under the e-book tab.

BACTERIOLOGY RESEARCH DEVELOPMENTS

BOTULINUM TOXIN

USES, MECHANISMS OF ACTION AND COMPLICATIONS

FAUSTO E. MARINO
AND
SEBASTIANO DE LUCA
EDITORS

NOVA BIOMEDICAL

New York

For permission to use material from this book please contact us:
Telephone 631-231-7269; Fax 631-231-8175
Web Site: http://www.novapublishers.com

NOTICE TO THE READER

Library of Congress Cataloging-in-Publication Data

ISBN: 978-1-62257-185-7

Library of Congress Control Number: 2012939702

Published by Nova Science Publishers, Inc. † New York

Contents

Preface		**vii**
Chapter I	Botox: Mechanism of Action, Uses And Side Effects *Penny Robshaw*	**1**
Chapter II	Effects of Botulinum Toxin on Neuronal Communication: Converging Evidences of Central Effects *C. Rasetti-Escargueil*	**37**
Chapter III	Aesthetic Indications of Botulinum Toxin: A Review of the Efficacy and Safetyof Currently Available Products *Luca A. Dessy, Nefer Fallico, Marco Mazzocchi and Nicolò Scuderi*	**53**
Chapter IV	The Use of Botulinum Toxin in the Treatment of Pain *Dominique Batifol and Pierre-Jacques Finiels*	**71**
Chapter V	Lower Facial Contouring with Botulinum Toxin: Functional and Aesthetic Aspects *Raffaele Rauso and Roberto Amore*	**87**
Chapter VI	Botulinum Toxin in Urology: Uses, Mechanisms of Action, Results and Complications *Cristobal Marchal, Felipe Saez-Barranquero and Maximino Redondo*	**103**
Chapter VII	Botulinum Toxin and Facial Palsy *M. L. Navarrete, L. Torrent, J. Junyent, J. F. Fuentes, J. R. Moya and A. L. Arellano*	**113**
Chapter VIII	Treatment of Body Deformities Due to Muscle Hypertrophy with Botulinum Toxin A *Luca A. Dessy, Marco Mazzocchi, Nefer Fallico and Nicolò Scuderi*	**125**

Chapter IX Effectiveness of Botulinum Toxin Type A Injection for TMJ Pain and Clicking **135**
Ali Hassani, Mohammad Hosein Kalantar Motamedi, Sarang Saadat and Mohammad Hamed Mansouri

Chapter X Botulinum Toxin A for Drug-Refractory Trigeminal Neuralgia **143**
Behnam Bohluli and Mohammad Hosein Kalantar Motamedi

Index **149**

Preface

Botulinum toxin (BoNT) is a powerful neurotoxin produced by the bacterium Clostridium botulinum. Its mechanism of action consists of the induction of a temporary muscle paralysis through the inhibition of the release of acetylcholine into the synaptic cleft. In this book, the authors discuss the uses and complications of botulinum toxin. Topics include the potential therapeutic applications of botulinum toxin in pain management and human epilepsy; a review of the efficacy and safety of currently available products; lower facial contouring with botulinum toxin; BoNT in urology, facial palsy, treatment of body deformities due to muscle hypertrophy, and for drug-refractory trigeminal neuralgia.

Chapter I - Botulinum neurotoxin type A (BoNT-A) is increasingly used in clinical practice for pharmaceutical and cosmetic purposes. It can also be used to treat a wide spectrum of conditions affecting the muscles and more recently, BoNT-A has been introduced as a new, minimally invasive treatment for both neurogenic and idiopathic detrusor overactivity (DO) along with treating pain due to pelvic floor muscle spasm.

Despite the fact that botulinum toxin (BTx) has been around for centuries, there is still a lack of evidence-based information that can guide clinical practice within the field of Urogynaecology. This chapter aims to look at the history of BTx, to discuss the authors' current understanding of its mechanisms of action and usage within the field of Urogynaecology and to consider what still needs to be done in order to increase our knowledge and improve clinical practice in the future.

Chapter II – Botulinum neurotoxins (BoNTs) are increasingly used thera-peutically to locally paralyse muscles for clinical or cosmetic benefit. Initially used to treat strabismus in the 1970s, botulinum toxin has more than a hundred possible medical applications: strabismus, movement disorders (dystonia and muscle hyperactivity), including hemifacial spasm, essential tremor, tics, cervical dystonia, cerebral palsy and vascular cerebral stroke. This deadly poison traditionally considered as a paralytic agent has soon been reported to provide pain relief prior to spasm relief. Despite recurrent clinical evidences and now approved use in chronic migraine there is only limited experimental evidence supporting botulinum toxin action on the central nervous system. Understanding the mechanism of action related to the use of BoNTs in the treatment of pain is still desired. This chapter will provide a more detailed understanding of current knowledge relating to the central actions of botulinum toxins, with particular emphasis on its potential therapeutic applications to pain management and human epilepsy.

Chapter III - Botulinum toxin (BTX) is a powerful neurotoxin produced by the bacterium Clostridium botulinum. Its mechanism of action consists of the induction of a temporary muscle paralysis through the inhibition of the release of acetylcholine into the synaptic cleft. The finding that BTX smoothed facial lines when used therapeutically prompt researchers to study the toxin effect on glabellar lines. In fact, the onset of glabellar lines is closely related to the underlying muscles' activity. Nowadays, BTX injections for glabellar lines are commonly performed and an increasing number of formulations is becoming available.

This chapter provides a comparative evaluation of currently available botulinum toxins for the treatment of glabellar lines. Toxins have been divided into three groups: those that have obtained approval for cosmetic use (approved treatments), those that have not (off-label treatments), and those that are still under investigation (experimental treatments). The basic similarities and differences between available botulinum toxins are examined, and the pharmacokinetics and dosing are summarized.

Approved treatments are composed of a sole serotype of botulinum toxin. Indeed BTX type A (BTX-A) is the only approved for the treatment of glabellar lines. The main difference between BTX-A formulations is their molecular weight: original formulations are made of a BTX-A complex of 900 kDa while the latest is a 150 kDa BTX-A free from complexing proteins. With reference to the dosage, the main aspect emerging from this comparison is that despite considering the same serotype each formulation possesses distinctive characteristics due to its purification and manufacturing processes. As such, BTX formulations are not interchangeable. The methods of evaluation used for each toxin are investigated and the safety and efficacy data reported in the literature are reviewed. Major adverse events are uncommon while minor adverse events, such as headache, blepharoptosis, and injection-site reactions, are relatively frequent but transient.

Off-label treatments consist of botulinum toxins that are approved for therapeutic applications but are used off-label for cosmetic indications. Currently, the only formulation available on the market is BTX type B. Published evidence on its safety and efficacy is extremely scarce.

Experimental treatments are composed of BTX products under investigation or recently launched for aesthetic applications. These formulations are designed to overcome some of the limitations present in the older generation of products. However, some of these toxins may be easily purchased via the Internet but do not have any license neither for cosmetic nor therapeutic indications. For this reason, such products are potentially a severe health risk for patients.

Chapter IV - The widespread use of Botulinum toxin (BT) is nowadays well accepted in the treatment of pain of spastic origin and also for aesthetic purposes. In addition, a more accurate knowledge of its action has lead to new methods of research which seem to show new indications, many of them totally unknown, even in the specialized medical community.

The aim of this work is to present the authors' experience and first results in such cases of non-conventional indications.

Chapter V - To achieve a slim and smooth contour line of the lower face, conventional treatment has concentrated on two approaches: surgical resection of the masseteric muscle or modelling ostectomy of the square-angled jaw. Modelling ostectomy of a square-angled jaw is mandatory when jaw's angles hyperostosis is present. When a square-angled face is due only to masseter muscle hypertrophy, our attention can be focused only on soft tissue remodelling. The resection of the masseter muscle is likely to cause a variety of side effects such as bleeding, hematoma, facial nerve injury, and asymmetric resection. Although the operation itself is relatively simple, it is difficult to predict the outcome.

Botulinum toxin injections into the masseter muscle can be used as an alternative, non invasive treatment, for the contouring of the lower face, however it is not associated with permanent results.

Masseteric hypertrophy is known as the asymptomatic enlargement of the masseter muscle and is commonly associated with abnormal habits such as nocturnal bruxism and daytime clenching.

The mechanism involved in the treatment of masseteric hypertrophy with botulinum toxin injection are somewhat different from those for treating facial hyperkinetic wrinkles. An intramuscular injection of botulinum toxin leads to temporary partial denervation at the area of the neuromuscular synapses and eventually causes muscle atrophy.

Occlusal parafunction, such as tooth grinding and tooth clenching, are unvoluntary mainly nocturnal habits that sometimes result in an hypertrophy of masseter muscle. Otherwise, masseteric muscle hypertrophy aetiology seems to be correlated to congenital genesis, psycological factors, habits and occlusal disorders. Especially occlusal parafunction such as bruxism can be correlated with the genesis of masseteric muscle hypertrophy, because the more masticatory muscle are used the most their volume can grow-up. In clinical routinary practice occlusal parafunction are treated with occlusal devices or botulinum toxin injection.

Authors explain the role of botulinum toxin in the treatment of the square-angled face related to masseter muscle hypertrophy and the importance in treating occlusal disorders, if present, at the same time to prolong the contouring effect of the toxin injected.

Chapter VI - The authors analysed the use of botulinum toxin (BoNT) injections as a therapeutic alternative for patients diagnosed with lower urinary tract disorders, reviewing mechanisms of action, results and complications. For this purpose, the authors evaluated systematic reviews, controlled clinical trials and prospective observational studies in the following urological pathologies: benign prostatic hyperplasia (BPH), overactive bladder (OAB), neurogenic detrusor overactivity (NDO), vesico-sphincter dyssynergia (BND) and interstitial cystitis / painful bladder syndrome (IC / PBS). There is evidence of the effectiveness of treatment with BoNTA in all or some of the symptoms and signs of the OAB, NDO, BND and IC/PBS. However, in BPH there is no clear evidence of the usefulness of this treatment. The main side effect observed was urinary retention.

Chapter VII - The therapeutic indication of peripheral facial paralysis depends on the degree of nerve injury. A severe facial palsy (electroneuronografic study less than or equal to 10%) develops into a healing sequelae. The sequelae of facial paralysis are the contractures, hemifacial spasm and synkinesis.

The authors' experience demonstrates that these patients may benefit from rehabilitation treatment (facial exercices and botulinum toxin). The authors present their work with patients affected by severe peripheral facial paralysis.

They were treated from the beginning of reinnervation by facial exercises according to the Wisconsin School and botulinum toxin. The authors observe that the subjective and objective efficacy of rehabilitation is high.

Rehabilitation treatment can inform the patient about his chances of recovery, take control and quality of facial expression, greater facial symmetry and thus provide better functionality and quality of life.

Chapter VIII - As cosmetic therapy, botulinum toxin type A (BTX-A) received Food and Drug Administration approval for the treatment of glabellar wrinkles in 2002. Besides this approach, many proven off-label indications of BTX-A are described in the literature.

Patients who practice intense sport activity may develop body deformities ascribable to muscle hypertrophy such as pseudogynecomastia due to the hypertrophy of the pectoralis major muscles, or prominent buttock due to the hypertrophy of gluteus muscles. However, classifications and treatment methods do not take into account such deformities. These forms are particularly evident in body builders with muscles progressively enlarging due to hypertrophy of muscle fibers, but may also be evident in slender individuals practicing non-agonistic sports. In most cases, this aspect is desired and appreciated, but it may be unpleasant for some individuals (i.e. breast with a female appearance or prominent buttock in males). In such cases, consultation with a plastic surgeon may be requested in order to determine the actual presence and nature of the deformity. If it is confirmed that the complained deformity is related to a specific muscle group hypertrophy, a possible solution of the problem is to reduce muscle activity. When a rapid reduction of muscular hypertrophy is requested by the patient, BTX-A injections may represent an option, being an effective non-surgical treatment modality.

Injections of BTX-A are a minimally invasive, very well-tolerated procedure for the treatment of hyperkinetic wrinkles and many other pathologies. The rationale for its use in such deformities consists in the induction of a temporary muscle paralysis by inhibiting the release of acetylcholine at the neuromuscular junction. Usually, one to two sessions of BTX-A injections are sufficient in order to reduce muscle hypertrophy and to obtain the desired result.

In conclusion, BTX-A injections are feasible, cost-effective and repeatable. Furthermore, this treatment achieves evident results with good patient satisfaction, and compliance. The proposed off-label indication should be considered when a rapid resolution of body deformities due to muscle hypertrophy is requested by the patient.

Chapter IX - *Background and Objective*: Temporomandibular joint (TMJ) disorders are among the common causes of head and neck pain; this is also associated with articular crepitus, muscle spasm and movement limitation. Various treatment methods like pharmaceutical therapy, occlusal adjustment, splint therapy and eventually surgery have been used for these patients each having their own advantages and disadvantages. New methods are sought for fast, conservative and complication-free treatment where other treatment modalities fail. At present, use of Botulinum toxin as a new treatment method is under investigation.

Materials and Methods: This prospective interventional study was conducted on 25 patients including 12 females and 13 males suffering from unilateral or bilateral TMJ pain and clicking. These patients had previously received 3 months of anterior repositioning splint therapy with no success. All patients were informed about the treatment with Botulinum toxin and filled out the consent form before initiation of treatment. The study was approved by the authors' ethics committee. Botulinum toxin A was injected into the masseter and temporalis muscles. Volume of injection was 25 units for the temporalis and 50 units for the masseter muscle 5-8 mm deep at three points (anterior, middle and posterior in the temporalis muscle and posterior deep, anterior superficial and inferior in the masseter muscle).

Results: Findings demonstrated that pain scores after 2 weeks and 8 weeks of follow up were decreased it but they did not show a significant reduction (P value=0.1). Minimal changes were detected in terms of clicking (P value=0.1) as well.

Conclusion: In general, this study showed that injection of BTX-A into masseter and temporalis muscles was minimally effective for improvement of clicking and pain.

Chapter X - Botulinum toxin type A (BTX-A) can be used to treat migraine and occipital neuralgia. BTX-A can also be used on trigeminal neuralgia (TN) patients who are refractory to medical treatment. Patients suffering from drug-refractory TN have been treated. The authors inject 50 U reconstituted BTX-A solution at the trigger zones. Symptoms, including pain duration, provoking factors, affected nerve branch, frequency of TN attacks, and severity of pain just before injections are evaluated 1 week, 1 month, and 6 months after injection. The overall response to treatment compared with values at baseline show men and women aged 28-67 years (mean 48.9 y) suffering from TN from 6 months to 24 years improve regarding frequency and severity of pain attacks; in the authors' patients, pain was completely eradicated and there was no need for further medication. In some patients treated, nonsteroidal antiinflammatory drugs are enough to alleviate pain attacks, and some patients again respond to anticonvulsive drugs after injection. All patients develop higher pain thresholds after injections. A significant difference in frequency of attacks before injection and at 1 week, 1 month, and 6 months after injection is seen. Significant improvement in all patients up to 6 months after injection can be anticipated. Complications include transient paresis of the buccal or mandibular branch of the facial nerve . BTX-A is a minimally invasive method that can play a role in treating TN before resorting to other more invasive therapies, i.e., radiofrequency and surgery.

In: Botulinum Toxin
Editors: Fausto E. Marino and Sebastiano De Luca
ISBN: 978-1-62257-185-7

Chapter I

Botox: Mechanism of Action, Uses and Side Effects

Penny Robshaw
MRCOG, Urogynaecology Department, St Mary's Hospital,
London, UK

Introduction

Botulinum neurotoxin type A (BoNT-A) is increasingly used in clinical practice for pharmaceutical and cosmetic purposes. It can also be used to treat a wide spectrum of conditions affecting the muscles [1] and more recently, BoNT-A has been introduced as a new, minimally invasive treatment for both neurogenic and idiopathic detrusor overactivity (DO) along with treating pain due to pelvic floor muscle spasm [2].

Despite the fact that botulinum toxin (BTx) has been around for centuries, there is still a lack of evidence-based information that can guide clinical practice within the field of Urogynaecology. This chapter aims to look at the history of BTx, to discuss our current understanding of its mechanisms of action and usage within the field of Urogynaecology and to consider what still needs to be done in order to increase our knowledge and improve clinical practice in the future.

History of Botox

The term "Botulinum" comes from the Latin word, *botulus*, which literally means "sausage". BTx was originally termed the "sausage protein" due to its association with meats that were improperly prepared, for example, sausages. At the end of the 18th century, a well-documented outbreak of "sausage poisoning" in Southern Germany, especially in the town of Württemberg, prompted early systematic botulinum toxin research. The German poet and district medical officer, Justinus Kerner, published the first accurate and complete descriptions of the symptoms of food-borne botulism between 1817 and 1822 [3]. Kerner did not succeed in defining the suspected "biological poison" which he called the "sausage poison" or fatty poison". However, he developed the idea of a possible therapeutic use of the toxin by noting that small amounts of the "sausage poison" might be useful for neurological conditions, such as "St Vitus Dance", more commonly known today as Sydenham's Chorea [3].

In 1895, eighty years after Kerner's work, there was a botulism outbreak after a funeral dinner of smoked ham in the small Belgium village of Ellezelles. Dr Emile Van Ermengem, Professor of Bacteriology at the University of Ghent, discovered the connection between the botulism outbreak and the bacterium, *Clostridium botulinum* [4]. The next major breakthrough came in 1928 when P Snipe and Hermann Sommer purified the toxin for the first time [5]. This was followed in 1949 by Arnold Bergen's discovery that botulinum toxin blocks neuromuscular transmission through acetylcholine release [6]. In the late 1960s, Alan Scott, a San Francisco ophthalmologist and Edward Schantz were the first to work on a standardized botulinum toxin preparation for therapeutic purposes. In 1973, Scott used BoNT-A in monkey experiments and followed this in 1980 with its first use in humans for the treatment of strabismus and blepharospasm [5].

The cosmetic effect of BoNT-A on wrinkles was documented in 1989 by a plastic surgeon from California, Dr Richard Clark, and was published in the Journal of Plastic and Reconstructive Surgery [7]. The same year, the United States Food and Drug Administration (FDA) approved the use of BTx for the treatment of strabismus and Blepharospasm. This was followed in 1992, by the publication of a study showing the safety of BoNT-A in the treatment of glabellar frown lines by husband and wife, Dr Alastair and Jean Carruthers [8]. In 1993, Pasricha's group showed that BTx could be used for the treatment of achalasia, a spasm of the lower oesophageal sphincter [9] and a year later, Bushara showed that BTx injections inhibit sweating [10]. This latter finding was the first demonstration of how BTx has a non-muscular use.

BTx has been used "off label" in the bladder for over two decades. The earliest published report of the use of botulinum toxin for urinary symptoms was the successful treatment in 1988 of detrusor sphincter dyssynergia in patients with spinal cord injury [11]. Schurch's group in 2000, were the first to describe injecting BoNT-A into the detrusor muscle of the human bladder for the treatment of neurogenic incontinence [12]. Although a randomised, placebo-controlled study (2005) in patients with neurogenic detrusor overactivity (NDO) demonstrated the efficacy of BoNT-A [13], as have open label studies [14] [15], it was not until 2011 that the FDA finally granted approval for the use of Onabotulinumtoxin A (Botox®, Allergan Inc.) for the treatment of urinary incontinence resulting from detrusor overactivity associated with a neurological condition in adults who have an inadequate response to or are intolerant of an anti-cholinergic medication. To date, the FDA has approved seven medical uses of BoNT-A (table 1).

Table 1. FDA approved medical uses of botulinum toxin

Date	Medical use
1989	Strabismus and Blepharospasm
2000	Cervical dystonia in adults
2002	Frown (glabellar) lines
2004	Severe primary axillary hyperhidrosis (*excess sweating*)
2010	Increased muscle stiffness in the elbow, wrist, fingers and muscles with upper limb spasticity
2010	Chronic Migraine (*defined as experiencing a headache most days of the month*)
2011	Urinary incontinence resulting from DO associated with a neurological condition in adults who have an inadequate response to or are intolerant of an anti-cholinergic medication

Further placebo-controlled, blinded studies must be completed, and the safety and efficacy determined, before BoNT-A can be approved for the treatment of other lower urinary tract symptoms. Until then, BoNT-A use will continue to be "off label".

Table 2. Clinical features of Botulism (*WHO fact sheet N°270, 2002*)

Botulism	
Food Sources:	Food sources of botulism include fermented, salted or smoked fish and improperly preserved home-processed canned or bottled foods with low acid content eg.Vegetables and meats. Commercial heat pasteurization may not be sufficient to kill all spores.
Inactivation:	Normal cooking processes – heating at >85°C for 5mins or boiling for a few mins
Incubation period:	Usually 18-36 hours (min: 4 hours; max: 8 days)
Symptoms:	The characteristic symptoms and signs are marked fatigue, weakness and vertigo, usually followed by blurred vision, dry mouth and difficulty swallowing and speaking. Vomiting, diarrhoea, constipation and abdominal swelling may occur. The disease can progress to weakness in the neck and arms, after which the respiratory muscles and muscles of the lower body are affected. The paralysis may make breathing difficult. There is no fever and no loss of consciousness.
Treatment:	Antitoxin administration is indicated as soon as possible after the clinical diagnosis has been made. Severe botulism cases require supportive treatment, especially mechanical ventilation, which may be required for weeks or months. Antibiotics are not required.
Prevention:	Prevention is based upon good hygiene and food preparation techniques. There is a vaccine against botulism, however, it is rarely used as its effectiveness has not been fully evaluated. Food and water samples associated with suspect cases must be obtained immediately, stored in properly sealed containers and sent to reference labs in order to help prevent further cases.
Mortality rate:	5-10% (if treatment does not start immediately)

Botulinum Toxin

Botulinum toxins are neurotoxic proteins produced by the bacterium, *Clostridium botulinum* (and rarely by other Clostridium species such as *C.butyricum*, *C. baratii* and *C. argentinense*), a spore-forming, anaerobic, gram-positive bacillus, that is found in soils and ocean sediment. Seven serotypes exist, A, B, C [C1 and C2], D, E, F and G, that are structurally similar, but antigenitically and serologically different. Botulinum toxin is responsible for causing botulism (table 2), a rare, but very serious disease, with a potential mortality rate of 5-10% if not treated immediately. The clinical syndrome of botulism can occur following ingestion of contaminated food, from colonization of the infant gastro-intestinal (GI) tract or from a wound infection. Human botulism is caused mainly by types A, B, E and (rarely) F. Serotypes C and D cause toxicity only in animals.

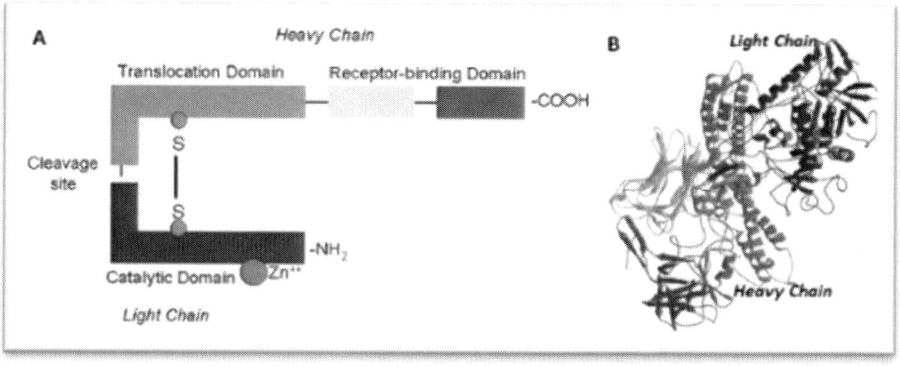

Figure 1.Botulinum toxin structure. A) Schematic representation of botulinum toxin. B) Crystal structure of botulinum toxin. The light chain catalytic domain is coloured blue. The heavy chain translocation domain is coloured green and the N-terminal and the C-terminal receptor binding domains are coloured in yellow and red respectively. The catalytic zinc is represented in grey. The colour code is the same for figures 1A and 1B [16].

Botulinum toxin is expressed as a single polypeptide chain (150kDa) which is activated by proteolytic cleavage into a two-chain polypeptide with a 100kDa heavy chain joined by a disulfide bond to a 50kDa light chain (figure 1).The heavy chain comprises of a translocation domain and a receptor-binding domain. The light chain (catalytic domain) is a zinc-containing metalloproteinase. Botulinum toxin causes flaccid muscle paralysis as opposed to the spastic paralysis seen in tetanus. It is probably the most acutely toxic substance known, with a median lethal dose of approximately 1ng/kg IV and 3ng/kg inhaled. Depending on the method of introduction, 90-270ng of botulinum toxin could be enough to kill an average 90kg person and 4kg, evenly distributed, could kill the entire human population of the world. Despite this toxicity, BTx has become a powerful therapeutic tool in the treatment of many medical conditions.

Mechanism of Action of Botulinum Toxin

In 1949, Arnold Burgen's group discovered that botulinum toxin blocks neuromuscular transmission via decreased acetylcholine (ACh) release [6]. Inhibition of ACh has been extensively investigated and is now largely accepted as the primary effect of botulinum toxin, however, there are now increasing reports of further mechanisms of action. Several studies suggest that botulinum toxin also acts on afferent neurotransmission by both inhibiting other neurotransmitters and by reducing the expression of afferent receptors [17].

Inhibition of Acetylcholine

Paralysis of the neuromuscular junction proceeds sequentially through a series of events. Firstly, Botulinum toxin binds highly selectively to cholinergic nerve endings, for example, synapses of the neuromuscular end-plate, sympathetic and parasympathetic ganglion cells and postganglionic parasympathetic cells [17]. The receptor-binding domain of the heavy chain (figure 1) provides cholinergic specificity and is responsible for binding the toxin to the pre-synaptic receptors [18]. The toxin then enters the cell via receptor-mediated endocytosis. The translocation domain of the heavy chain promotes translocation of the light chain across the endosomal membrane [18].

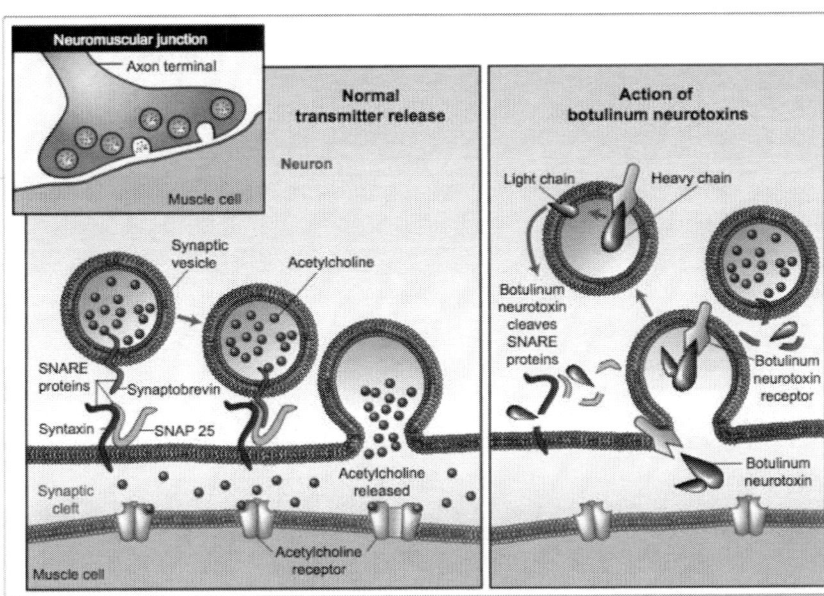

Figure 2. Mechanism of action of botulinum toxin (right) compared to normal cell (left). Shown are the individual stages of botulinum toxin intoxication, including cell surface recognition, vesicle internalization, translocation of the catalytic domain of the light chain into the cytosol and proteolytic cleavage of one or more proteins of the SNARE complex. These steps lead to inhibition of neurotransmitter vesicle release. Serotypes B, D, E and G cleave proteins of the VAMP family (blue) and serotypes A, C and E cleave SNAP-25 (yellow). Serotype C can also cleave Syntaxin (purple). Taken from Rowland [18].

In normal ACh release, the SNARE protein complex is responsible for vesicle fusion and release of ACh into the synapse. The SNARE protein complex is formed from several large proteins, including SNAP-25, Syntaxin and VAMP. The light chain acts as a zinc endopeptidase and causes proteolytic cleavage of one or more proteins of the SNARE complex, depending on the botulinum toxin serotype [18] [19]. Serotypes B, D, E and G cleave proteins of the VAMP family [18], A, C and E cleave SNAP-25 [18] [19] and serotype C also cleaves Syntaxin [18]. This action prevents formation of the SNARE complex and therefore, prevents fusion of the ACh vesicle with the pre-synaptic membrane resulting in the inhibition of ACh release and flaccid paralysis of the muscle. This action is shown below in figure 2.

Botulinum Toxin and its Action on Afferent Neurotransmission

Although the primary effect of Botulinum toxin A is the inhibition of ACh release, inhibition of other substances of the afferent limb of the micturition reflex (such as Adenosine 5'-triphosphate (ATP) and Calcitonin gene-related peptide (CGRP)) and reduction in the expression of afferent receptors (such as TRPV1) also contribute to the beneficial effect of BoNT-A in the urinary bladder [17].

ATP and Botulinum Toxin

ATP has an important signaling role in both the central and peripheral nervous system. It is recognized by purinergic receptors, such as P2X and P2Y. There is evidence that ATP could be a nonadrenergic-noncholinergic (NANC) mediator in human bladder smooth muscle. Purinergic transmission works on both afferent and efferent signaling pathways within the lower urinary tract and involves multiple ATP receptors including P2Y metabotropic and P2X ionotropic receptor families [19]. Immunohistochemical studies have shown ATP receptors (P2X subtypes) on detrusor muscle cells, bladder nerves and on urothelial cells.

ATP also appears to be involved in bladder sensation, and urothelial derived ATP release may have a role in both autocrine and paracrine signaling in the bladder [19]. Urothelial-derived ATP release during bladder distension could trigger sensations of fullness and pain or induce changes in bladder activity [20]. Inflammation within the bladder, along with P2X receptor agonists, has been shown to increase bladder afferent firing [21]. Kumar *et al.* reported a significant increase in ATP release from the urothelium of patients with painful bladder syndrome, suggesting that this type of noncholinergic mechanism may play a role in bladder pathology [22].

Khera *et al.* investigated ATP release in the bladders of rats with spinal cord injury (SCI) and in healthy rats. They reported that there was a greater release of ATP in those rats with SCI, which they believe contributed to the development of detrusor overactivity. However, intravesical injection of BoNT-A resulted in a significant inhibition of the increased ATP release [23]. This would reduce its proposed excitatory effect on suburothelial and urothelial purinergic receptors [24] and in turn, decrease bladder afferent firing. BoNT-A has also been shown to reduce detrusor overactivity evoked by intravesical adenosine triphosphate and capsaicin in rat bladders [25].

CGRP and Botulinum Toxin

CGRP is a member of one of the Calcitonin family of peptides and is one of the most abundant peptides produced in both peripheral and central neurons. It is a potent vasodilator and is thought to have a role in nociception and the transmission of pain. Studies have shown that women with idiopathic detrusor overactivity have increased levels of CGRP fibres compared to controls [24].

In a rat bladder pain model, intravesical BoNT-A administration produced significant improvement in the intervals between bladder contractions in parallel with a reduction of the release of CGRP from the superficial bladder layers [24]. In an acetic acid-induced bladder pain model in rats, Chuang *et al.* investigated the effect of intravesical BoNT-A on CGRP and bladder hyperactivity. They showed that intravesical BoNT-A administration blocks the induced bladder pain responses and inhibits CGRP release from the afferent nerve terminals [26]. Thus, it seems that inhibition of CGRP may contribute to the mechanism of action of BoNT-A within the bladder.

Afferent Receptors and Botulinum Toxin

Sensory afferent receptors, such as the transient receptor potential channel (TRPV1), have also been implicated in the mechanism of action of botulinum toxin [27]. This receptor is located predominantly in the membrane of capsaicin-sensitive C fibres and more recently, it has been detected in smaller amounts in the cell membrane of urothelial cells [17].

In normal health, the afferent pathway in the micturition reflex is mediated largely by A∂ fibres that send information about bladder fullness and wall tension to the pontine micturition centre via the periaqueductal gray matter [28]. However, after spinal disruption, a different type of afferent pathway emerges, mediated by capsaicin-sensitive C fibres. In these neurogenic patients, a spinal segmental reflex pathway is triggered, resulting in detrusor overactivity [29]. Evidence for this involvement of C fibres in the spinal reflex pathway, comes from reports of the successful treatment of patients with neurogenic detrusor overactivity with intravesical capsaicin, or its ultra potent analog, resiniferatoxin (RTX) [30] [31].

The TRPV1 receptor is a nonselective cation channel that plays a key role in pain perception and control of the micturition reflex [32], mainly during pathological conditions such as cystitis or spinal cord transection. Patients with neurogenic detrusor overactivity have substantially more TRPV1-receptor-positive nerve fibres than patients with normal bladders [33]. In patients with neurogenic DO of spinal origin, intravesical resiniferatoxin resulted in the decreased expression of TRPV1 in bladder sensory fibres [34] [35] and in urothelial cells [27], which correlated with an increase in bladder capacity and a decrease in urgency [36].

BoNT-A has been shown to prevent TRPV1 exocytosis to the plasma membrane during bladder inflammation [37]. As a result, the hypothesis is that the activation of the micturition reflex by bladder filling is prevented as well as noxious transmission within the bladder. Apostolidis *et al.* studied TRPV1, and the purinergic receptor P2X3, in bladder specimens of patients with both neurogenic and idiopathic DO and controls following BoNT-A injections

[27]. They found that there were significantly fewer P2X3-immunoreactive fibres and TRPV1-immunoreactive fibres 16 weeks after BoNT-A administration, and this correlated with an improvement in both clinical and urodynamic outcome [27]. This suggests that BoNT-A can mediate afferent sensory receptor expression and that this may contribute to its increasingly complex mechanism of action.

Substance P and Botulinum Toxin

Another possible component of the mechanism of action of BoNT-A is its interaction with substance P (SP), in particular, that BoNT-A inhibits its release from suburothelial nerves [24].

SP is a neuropeptide that functions as both a neurotransmitter and a neuromodulator. It is released from the terminals of specific sensory nerves and is associated with inflammatory processes and pain. There is documented evidence that SP shares a common pathway with ATP and ACh and activates calcium-dependent nonselective cation channels, including TRPV1 [38]. SP may also potentiate the susceptibility of purinergic receptors, such as P2X3, to ATP [39]. Inhibition of SP would therefore reduce the activation of TRPV1 and P2X3 receptors.

Lucioni *et al.* confirmed the hypothesis that BoNT-A inhibits SP release. They studied an inflammatory rat bladder model and showed that BoNT-A inhibited the release of both SP and CGRP [40]. Welch *et al.* studied the effects of different botulinum toxin subtypes on rat dorsal root ganglion neurons and showed that subtypes A, B, C and F all inhibited SP release, however, BoNT-A demonstrated the greatest potency for inhibition of the sensory neurons [41]. Of note, they also showed that the inhibition of SP by the BTx subtypes correlated with cleavage of their respective BTx substrate proteins (eg BoNT-A and SNAP-25) [41], lending weight to the hypothesis that inhibition of SP occurs through a similar mechanism to that observed when BoNT-A inhibits ACh release at the neuromuscular junction.

These findings suggest a possible use for BoNT-A in the treatment of neurogenic inflammation that may underlie bladder disorders such as interstitial cystitis. However at present, there remains conflicting evidence as to the effectiveness of this treatment.

Botulinum Toxin and Nerve Growth Factor (NGF)

Following the detection of high levels of NGF in the urine of patients with DO and chronic bladder inflammatory conditions [42], it has gained importance as a modulator of the micturition reflex and as a potential therapeutic target.

NGF is a small, secreted protein that in the urinary tract is produced by the urothelium and smooth muscle. It is critical for the survival and maintenance of sympathetic and sensory neurons, appears to have a role in tissue inflammation, and can sensitize afferent nerves and induce bladder hyperactivity [43].

Increased levels of NGF in urine have been shown to induce hyperactivity of the urinary bladder in rats [44]. Increased levels have also been demonstrated in women with idiopathic sensory urgency and interstitial cystitis [45]. Liu *et al.* (2009) confirmed this finding, showing that urinary NGF is elevated in patients with IDO and NDO compared to controls. Detrusor BoNT-A treatment reduced urinary NGF levels in responders with IDO and NDO [46]. This suggests that NGF is associated with activation of sensory receptors, such as TRPV1, resulting in overactive symptoms and that BoNT-A can mediate afferent sensory nerves by the inhibition of NGF.

Conclusion

The underlying pathophysiology of detrusor overactivity and interstitial cystitis is still yet to be fully determined. BoNT-A has been shown to have a complex mechanism of action. Although the primary peripheral effect of botulinum toxin is the inhibition of Ach, inhibition of other substances of the afferent limb of the micturition reflex and reduction in the expression of afferent receptors also contribute to its beneficial effect on the urinary bladder. These events at the molecular level translate into the therapeutic effects seen following BoNT-A injection.

Clinical Effects of Botulinum Toxin

Over the last decade, botulinum toxin has become a powerful therapeutic tool for the treatment of many medical conditions. Many of its uses in Urogynaecology are still "off-label", however it was granted FDA approval for the treatment of NDO in 2011. Although there are 7 serotypes of Botox, only types A and B are used in clinical practice. The different preparations have different doses, efficacy and safety profiles and should not be considered generic equivalents. The potency is expressed in units of activity. Table [3] lists those botulinum toxins currently marketed.

The majority of the commonly used botulinum toxins are serotype A and it is the clinical use of this subtype that this chapter will focus on. The use of serotype B will be discussed briefly at the end.

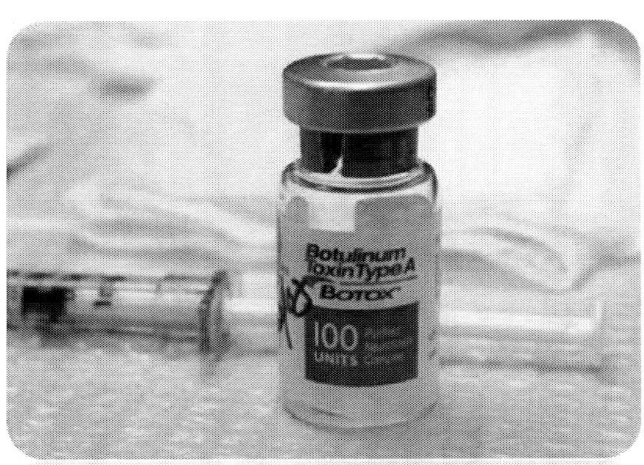

Figure 3. 100U vial Botox®

Table 3. Preparations of botulinum toxin currently on the market

Serotype	A	A	A	B
Generic name	Onabotulinumtoxin A	Abobotulinumtoxin A	Incobotulinumtoxin A	Rimabotulinumtoxin B
Brand name	Botox	Dysport	Xeomin	Myobloc/Neurobloc
Manufacturer	Allergan Inc. (US)	Ipsen (France)	Merz pharmaceuticals GmbH (Germany)	US World Meds
Packaging (U/vial)	100	500	100	2,500 to 5,000/10,000
Preparation	Dry; vacuum dried	Dry; lyophilized	Dry; lyophilized	Solution (5000 U/ml)
Storage of packaged product	-5°C or 2-8°C	Room temperature	Room temperature	2-8°C
Storage after reconstitution	2-8°C for 24 hours	2-8°C for several hours	2-8°C for 24 hours	Few hours
Specific activity (μ/ng)	20	40	167	75-125

The use of Botox in cosmetic procedures is well recognized, however, Botox can be used to treat a wide spectrum of conditions affecting the muscles, ranging from spastic muscle disorders to muscle under activity, such as detrusor hypotonia. Within urology, Botox has been used in the treatment of detrusor external sphincter dyssynergia, detrusor hyperreflexia, the non-relaxing urethral sphincter, chronic prostate pain, pelvic floor muscle spasticity, interstitial cystitis, non-fibrotic bladder outflow obstruction (including benign prostatic hyperplasia (BPH)), detrusor overactivity, detrusor underactivity and acute urinary retention in women. Table [4] highlights the many uses of Botulinum toxin within medicine, with those conditions highlighted in green being the only ones that have received FDA approval. Medical practitioners continue to use Botox "off label" for the remainder.

It is impossible to discuss all of the possible uses of botulinum toxin within Urogynaecology in detail and so the focus of discussion will be on the clinical effects of BoNT-A on four major conditions: Detrusor sphincter dyssynergia, detrusor overactivity, interstitial cystitis/painful bladder and pelvic floor muscle pain/spasm.

Detrusor Sphincter Dyssynergia

Detrusor sphincter dyssynergia (DSD) is defined as an involuntary contraction of the external urethral sphincter during a detrusor contraction [47]. This is invariably a consequence of neurological pathology, such as spinal cord injury (SCI) or multiple sclerosis, and results in increased intravesical pressures, voiding dysfunction and high residual volumes.

Table 4. Uses of Botulinum Toxin

Focal dystonias: involuntary, sustained or spasmodic patterned muscle activity
Cervical dystonia (spasmodic torticollis)
Blepharospasm (eyelid closure)
Laryngeal dystonia (spasmodic dysphonia)
Limb dystonia (writer's cramp)
Oromandibular dystonia
Orolingual dystonia
Truncal dystonia
Spasticity: velocity-dependent increase in muscle tone
Stroke
Traumatic brain injury
Cerebral palsy
Spinal cord injury
Non-dystonic disorders of involuntary muscle activity
Hemifacial spasm
Tremor
Tics
Myokymia and synkinesis
Myoclonus [tensor veli palatini muscle (middle ear)]
Hereditary muscle cramps
Strabismus and nystagmus
Disorders of localized muscle spasms and pain
Chronic low back pain
Myofascial pain syndrome
Temporomandibular joint disorders
Tension headache
Migraine headache
Cervicogenic headache
Smooth muscle hyperactive disorders
Achalasia cardia
Hirschsprung's disease
Sphincter of Oddi dysfunctions
Chronic anal fissure
Cosmetic use
Hyperkinetic facial lines (glabellar frown lines)
Hypertrophic platysma muscle bands
Sweating disorders
Axillary and palmar hyperhidrosis
Frey syndrome
Urological
Detrusor external sphincter dyssynergia
Detrusor hyperreflexia (NDO)
Spastic conditions of the urethral rhabdosphincter
Chronic pelvic pain (and pelvic floor spasticity)
Interstitial cystitis, motor and sensory urge incontinence
Non-fibrotic bladder outlet obstruction (and BPH)
Detrusor underactivity
Acute urinary retention in women

As a consequence, complications such as hydronephrosis and infection can occur. Patients with spinal cord injury are particularly vulnerable, with 67% of all deaths occurring as a result of renal failure, likely secondary to voiding dysfunction [48]. Standard treatments, such as spasmolytic agents or alpha-blockers, are often unsuccessful, whereas intermittent self-catheterization, although a convenient therapy, is not suitable, or technically possible, for many spinal cord patients. Sphincterotomy, for many years, was the only alternative treatment, however it is far from satisfactory as high re-operation rates and long-term clinical failures have been described [49].

Several studies have evaluated the efficacy of BoNT-A in treating DSD. Varying doses were injected either transurethrally [50] [51] into the external urethral sphincter using a rigid cystoscope or transperineally [48] [52]. Outcomes were similar in all studies, with BoNT-A significantly reducing post-void residuals and decreasing voiding pressures [48] [50] [51] [52]. Patients reported significant subjective improvements in voiding and some were able to stop intermittent self-catheterization during the treatment.

Both methods of administering the botulinum toxin were effective, with no significant differences reported in terms of clinical outcome. Logic suggests that the transperineal approach will become preferable, as it is less invasive and does not require specialist equipment, additional staff or theatre time. Most studies used a dose of 100U with good results, and saw clinical benefits last, on average, 3 months [48] [51].

Botulinum toxin appears to be a safe and effective treatment for DSD in patients with spinal cord injury. Studies have shown that it improves voiding dysfunction, both clinically and urodynamically, with few side effects. The next step is to perform a randomised, placebo-controlled study to determine whether botulinum toxin has a future in the treatment of DSD.

Detrusor Overactivity

The International Continence Society (ICS) defines the overactive bladder syndrome as a syndrome characterized by urinary urgency, with or without urge incontinence, usually with frequency and nocturia, in the absence of proven infection or other obvious pathology [53] The term detrusor overactivity (DO) was adopted by the ICS as the corresponding urodynamic observation, characterized by involuntary detrusor contractions during the filling phase that may be spontaneous or provoked [53]. DO may be idiopathic, where no cause is identified, or neurogenic, resulting from an underlying neurological condition (eg. SCI or MS). Overactive symptoms are common affecting over 12% of the population above 18 [54], and have a significant impact on one's quality of life. Standard treatment involves bladder retraining and antimuscarinics; however, some do not respond to conventional therapy or find the side effects of the medication too troublesome.

Treatment options for this group include botulinum toxin A, sacral neuromodulation and augmentation cystoplasty. Of these, intravesical BoNT-A is becoming popular amongst both clinicians and patients, although its use still continues to be "off-label".

Studies in Neurogenic Detrusor Overactivity

Initial studies on the use of BoNT-A in detrusor overactivity were done in patients with neurogenic DO who either did not respond to, or tolerate, antimuscarinics. Prior to the advent of BoNT-A, these patients would require surgical treatment in the form of augmentation cystoplasty. The first published study on the use of BoNT-A as a treatment for neurogenic DO was by Schurch *et al.* in 2000 [55]. They studied 19 spinal cord injured patients with NDO and found that there was significant improvement in reflex volume (volume at first involuntary contraction), mean cystometric capacity (MCC), compliance and pdetmax, 12 weeks following treatment [55]. This was sustained for up to 36 weeks in some patients.

Following this study, researchers around the world started studying the therapeutic benefits of BoNT-A in DO. Many open-label studies, and some randomised controlled trials, were performed, usually in women with either spinal cord injury or multiple sclerosis. In 2004, Reitz *et al.* reported on a 200 patient, multicentre, open-label study on BoNT-A treatment of NDO [14]. Similar to the Schurch study, they injected 300U of BoNT-A at 30 sites, excluding the trigone, at a dilution of 100U/10mls. Their results showed significant improvement in all urodynamic parameters. At the initial follow-up (12 weeks) and the second follow-up (36 weeks) the MCC and the mean voiding pressure increased significantly. The mean bladder compliance increased significantly by the first follow-up examination, however, the increase was not significant at 36 weeks. In addition, 73% of incontinent patients became continent at 12 weeks, and in 72% of these, complete continence persisted at 36 weeks [14]. No injection-related complications or toxin-related side effects were reported.

In 2004, Giannantoni *et al.* reported results from the first controlled study of BoNT-A and NDO [56]. Patients were randomised to receive either 300U BoNT-A or 0.6μmol/L of intravesical resiniferatoxin (RTX). Repeat injections or installations were allowed. Both treatments resulted in improvements in continence and urodynamic parameters at 6, 12 and 18 months, however, the improvements were significantly better with BoNT-A for all variables at all time points [56]. In addition, RTX patients required more repeat installations (mean: 8.6) compared to BoNT-A (mean: 2.1).

The first randomized, placebo-controlled study was reported in 2005. Schurch *et al.* randomised 59 patients with NDO to receive either 200U BoNT-A, 300U BoNT-A or placebo [57]. Significant improvements in key urodynamic parameters were seen at all time points in the BoNT-A treated patients. After treatment, there were significant decreases in incontinence episodes at most time points in the two BoNT-A groups. The mean reduction in incontinence episodes for the two BoNT-A groups at all time points was approximately 50% [57]. There were also significant improvements in incontinence quality of life scores for both BoNT-A groups. The 200U dose seemed to be equally as effective as the 300U dose; however, no conclusions can be drawn from this, as the study was not powered to compare different doses.

In 2011, Herschorn *et al.* performed a multicentre, randomised placebo-controlled trial looking at the effectiveness of BoNT-A in 57 patients with a neurogenic bladder [58]. They showed that six weeks following injection, patients receiving BoNT-A had significantly fewer daily episodes of urinary incontinence. Urodynamic studies showed that both bladder compliance and MCC improved in patients receiving BoNT-A compared to placebo. Quality of life scores also improved significantly in the BoNT-A group [58]. These findings persisted for at least 24 weeks.

Despite heterogeneous study designs, almost all single-injection studies showed significant improvements, at the last follow-up, in outcomes such as incontinence episodes (mean reduction: 69%; range: 32-100%), maximum cystometric capacity (mean increase: 85%; range: 11-303%) and maximum detrusor pressure (mean decrease: 44%; range: 5-83%) [59]. Overall, the mean percentage of spinal NDO patients who became fully continent was 56% (range: 30-87%), however, return to full continence was only 8% in those patients with NDO due to a cerebrovascular accident [59].

To summarize, studies have shown that BoNT-A is a safe and effective treatment for the treatment of refractory NDO. As a result of this research, in 2011 the FDA finally granted approval for the use of BoNT-A in "urinary incontinence resulting from DO associated with a neurological condition in adults who have an inadequate response to or are intolerant of an anti-cholinergic medication".

Studies in Idiopathic Detrusor Overactivity

Through research into NDO, BoNT-A has been shown to be a promising treatment for refractory DO. The question remains as to whether it works equally well in DO of non-neurogenic origin.

The first published studies on IDO contained patients with both NDO and IDO. In 2004, Kuo investigated the urodynamic changes after detrusor injection of 200U BoNT-A in 30 patients (men and women) with refractory DO (both NDO and IDO) [60]. After BoNT-A injection, improvement in both urodynamic and clinical parameters were noted. The urodynamic results showed increased MCC, decreased voiding pressure and increased bladder neck opening time. Increased post-void residual volume and decreased voiding efficiency were also noted, however, these effects were temporary [60]. 26% of patients regained continence and 47% of patients noted an improvement in overactive symptoms. Following this study, Kuo concluded that BoNT-A is effective in the treatment of refractory DO [60].

Schmid *et al.* conducted the largest study on 100 patients (men and women) with idiopathic DO [61]. 100U of BoNT-A was injected at 30 sites, sparing the trigone. They found that urgency improved in 66% of patients at 12 weeks as well as frequency of urine being significantly reduced. 80% of incontinent patients reported complete continence at 12 weeks. Improvements were also seen in urodynamic parameters, with DO resolving in 74% of patients. Temporary urinary retention was seen in 4% of patients [61].

Several randomised controlled trials now exist that compare the efficacy of 200U BoNT-A against placebo [62] [63] [64]. Sahai *et al.* specifically looked at the effect on patients' quality of life (QoL) following BoNT-A injection [62]. QoL was assessed using the King's Health Questionnaire (KHQ). They found that BoNT-A injections improved the QoL in patients for at least 24 weeks. In particular, "Incontinence Impact", "Emotions", "Physical Limitations", "Social limitations" and "symptom severity" were significantly improved. Flynn *et al*, in 2009, found that there was a statistically significant decrease in pad use in those patients receiving BoNT-A, however, they found no difference in reported nocturia, frequency or detrusor pressure between the groups [63]. Brubaker *et al.* showed significant improvements in lower urinary tract symptoms in approximately 60% of women after intra-detrusor BoNT-A injection [64]. They did note a higher than expected rate of increased PVR (43%) and UTI (44%). One reason for this might have been their study design, where increased PVR was defined as a PVR>200mls regardless of symptoms or the ability of patients to void unassisted [64].

Intravesical injections of BoNT-A for the treatment of refractory IDO appears to significantly improve overactive symptoms, however, studies do highlight the possible risk of increased post-void residual and subsequent urinary retention. Providing patients are made aware of this risk, and are taught to perform CISC, then BoNT-remains an effective and safe treatment for refractory IDO. However, despite promising results from randomised trials, at present clinical use is still "off label".

Interstitial Cystitis/painful Bladder

Interstitial cystitis (IC) is a clinical diagnosis, primarily based on symptoms of urgency, frequency and bladder or pelvic pain. It mainly affects women and has an unclear aetiology. The ICS prefers to use the term painful bladder syndrome (PBS) and this will be the term used here. PBS is defined as "the complaint of suprapubic pain related to bladder filling, accompanied by other symptoms such as increased daytime and night-time frequency, in the absence of proven urinary infection or other obvious pathology" [53].

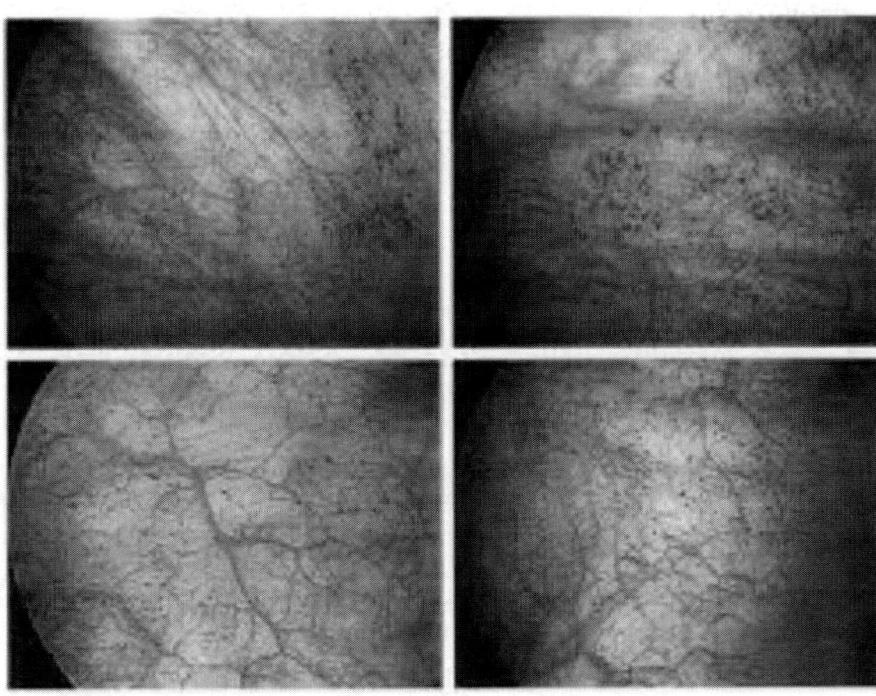

Figure 4. Severe petechial haemorrhages at cystoscopy

Similarly, the European Society for the study of IC (ESSIC) uses the term bladder pain syndrome, defined by pain in the bladder area increasing with bladder filling, urinary frequency and nocturia [65]. The ICS reserves the diagnosis of IC for patients who have "typical cystoscopic and histological features", although they do not go on to define what these are [53]. The National Institute of Diabetes and Digestive and Kidney disease (NIDDK) definition of IC requires findings of petechial haemorrhages or Hunner's ulcers at cystoscopy; however, the lack of both of these findings does not exclude the diagnosis of PBS.

There is a considerable lack of evidence supporting many of the treatments offered to patients with IC. IC is notoriously difficult to treat, often leaving both clinicians and patients frustrated. Standard treatments include behavioral modifications, oral medication and bladder instillations. Intravesical BoNT-A is emerging as a potential new treatment for refractory interstitial cystitis. The rationale behind this is the hypothesis that botulinum toxin has an antinociceptive effect on bladder afferent pathways.

Smith *et al.* treated 13 IC patients with 100U-200U of Dysport or Botox submucosally in the trigone and bladder base and found that 69% of patients reported subjective improvement after treatment: the symptom index improved by 71%, problem index by 69% and bladder pain by 79% [66]. They concluded that BoNT-A has an antinociceptive effect on bladder afferent pathways in patients with IC, producing both symptomatic and functional (ie. urodynamic) improvements [66].

In another study, Kuo used suburothelial injections of BoNT-A to treat 10 patients with IC [67]. Unfortunately, results were disappointing with improvements only being seen in 20%. In these 2 patients, there was a significant increase in functional bladder capacity and a significant decrease in frequency and pain score. Mild difficulty in urination was reported in 7 patients.

Giannantoni *et al.* treated 14 patients with injections of 200U BoNT-A into both the trigone and bladder base [68]. They reported subjective improvements in 85.7% of patients at both 1 and 3 months. In these patients, there was a decrease in scores on the visual analog scale along with a decrease in urinary frequency and a significant increase in bladder capacity. Two patients reported incomplete bladder emptying, however, there were no reported systemic side effects. Giannantoni *et al.* concluded from this that injections of intravesical BoNT-A were effective in the short-term management of PBS [68]. In a follow-up study, the same group noted that the effects of intravesical BoNT-A decreased progressively a few months after treatment, and therefore concluded that repeat injections would be needed in order for treatment to be effective and sustained [69].

All the studies examining the effects of BoNT-A in IC are small, with short-term follow-up. Outcomes are mixed, but the trend is still towards an improvement in symptoms following treatment. Intravesical BoNT-A may be of use in selected patients with refractory IC who have failed conventional treatments. Further randomized, controlled studies are needed before BoNT-A is recommended as a treatment for IC.

Pelvic Floor Muscle Spasm/Pain

Chronic pelvic pain is thought to affect 15% of women and is a cause of significant morbidity. Diagnosis can be difficult and managing these women is often challenging. Whilst for some, standard medical and surgical interventions are sufficient to treat symptoms; there remains a group of women for whom chronic pelvic pain is ineffectively treated. It is thought that pelvic floor muscle spasm can contribute to pelvic pain, however, it is not yet clear whether muscle spasm occurs secondary to pre-existing conditions, such as endometriosis, or whether primary pelvic floor myalgia can occur, like elsewhere in the body.

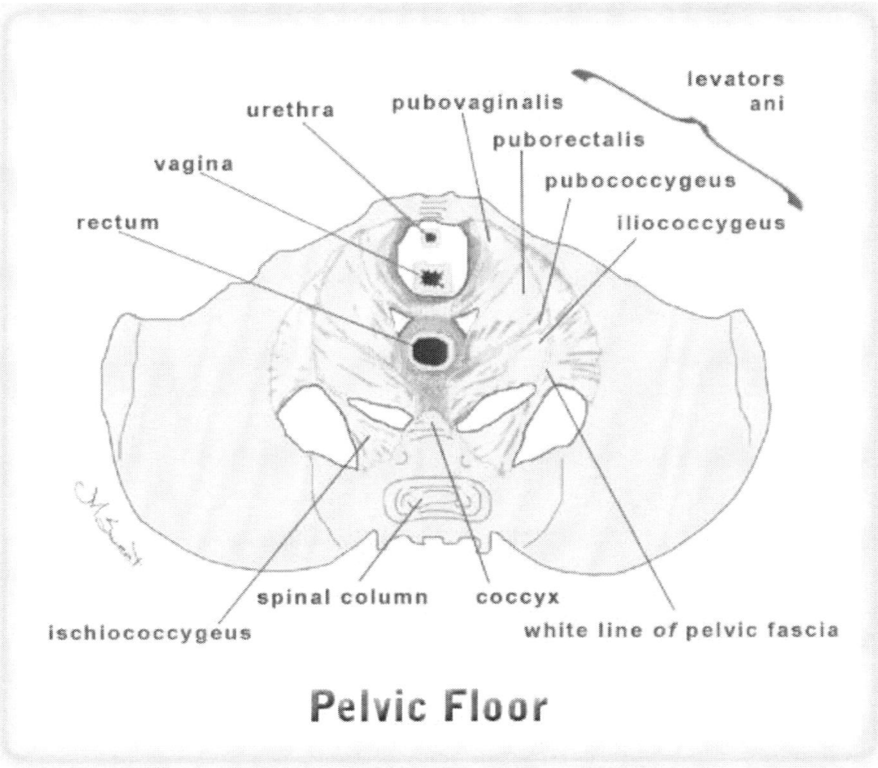

Figure 5. Diagram showing the muscles of the female pelvic floor.

If pain results from muscle spasm, then relieving that spasm should improve symptoms. Botulinum toxin is used in many medical disciplines as a treatment for muscle dystonias and spasticity, therefore, there is no reason why it should not be effective in treating pain due to pelvic floor muscle spasm. In fact, there is a surprising lack of published research into the use of BoNT-A for the treatment of pelvic floor muscle spasm. Most are case studies, or reviews of those case studies, however there is one randomised controlled trial. All studies deal with small numbers of women.

In 2005, Thomson *et al.* published a case study on a woman with PBS and chronic pelvic pain refractory to treatment [70]. She was known to have endometriosis, however, clinical examination revealed spasm of both puborectalis and pubococcygeus muscles. She received a total of 40U BoNT-A (10U at 4 sites) and following treatment, reported an improvement in bladder symptoms, pain, sexual activity and quality of life. They concluded from this that BoNT-A might be effective in reducing pain in women with hypertonicity of the pelvic floor muscles [70].

A year later, Brown *et al.* published data from a pilot study that examined the efficacy of BoNT-A injections into the levator ani muscles to relieve coital pain, reduce pelvic floor tension and instability and to reduce vestibular hyperalgesia in vestibulodynia [71]. Two patients with vestibulodynia were treated with 20U and 40U of BoNT-A. Pelvic floor hypertonicity and variability were markedly reduced in both patients, however, negligible changes occurred in vestibular hyperalgesia. Coital pain was improved in one patient only. They concluded that BoNT-A injections might be effective in reducing coital pain in vestibulodynia with levator ani tenderness [71].

The same year, Abbott *et al.* conducted a randomised, controlled, double blind, placebo controlled trial on 60 women with chronic pelvic pain and evidence of pelvic floor muscle spasm [72]. 30 women had 80U of BoNT-A injected into the pelvic floor muscles. There was a significant reduction in dyspareunia and non-menstrual pelvic pain in the BoNT-A group; however, there was also a significant reduction, albeit to a lesser degree, in the placebo group. Pelvic floor pressure (as measured by vaginal manometry) was significantly reduced in those women receiving BoNT-A, but again, lower pelvic floor muscle pressures were seen in the placebo group. They concluded that objective reduction of pelvic floor spasm reduces certain types of pelvic pain and that BoNT-A might be a useful agent in women with pelvic floor muscle spasm and chronic pelvic pain who do not respond to conservative physical therapy [72].

The few reports in the literature suggest that BoNT-A might be a potential therapeutic agent in the treatment of pelvic floor muscle spasm and chronic pelvic pain in select patients. The large placebo effect in Abbott's study highlights the difficulty that exists in determining what treatments are effective in the treatment of chronic pelvic pain. Larger controlled trials are warranted before BoNT-A can be marketed as a possible treatment for pelvic floor muscle spasm and chronic pelvic pain.

Vaginismus

One condition resulting solely from pelvic floor muscle spasm is Vaginismus. Vaginismus occurs when the muscles around the outer third of the vagina contract involuntarily when vaginal penetration is attempted. This makes penetrative intercourse painful or prevents it all together. It is probably under reported, however, studies have quoted an approximate prevalence of 8-13% [73].

BoNT-A was first described for the treatment of vaginismus in a case report published in *The Lancet* in 1997 [74]. A woman with vaginismus and urinary symptoms had 10U of BoNT-A injected into 2 sites on her anterior vaginal wall with good effect [74]. Shafik and El-Sibai described a study in which 8 women with vaginismus were injected with 25U of BoNT-A at multiple points into the bulbospongiosus muscle [75]. All 8 patients had complete resolution of their symptoms after 1 injection. Ghazizadeh and Nikzad reported the treatment of 24 women with refractory vaginismus with between 150-400U Dysport [76]. 75%

achieved satisfactory intercourse after the first injection and were still reporting improvements at 12 months.

Despite these promising results, BoNT-A should not be seen as a substitute for standard treatments, namely psychosexual counseling using psychotherapeutic or behavioral techniques. In the future, it may prove to be a useful adjunct to conventional treatments, however, far more randomised controlled studies are needed before this is the case.

Botulinum Toxin in Paediatric Urogynaecology

Botulinum Toxin has been used successfully in children for over a decade. The paediatric literature is focused on the use of BoNT-A in non-neurogenic and neurogenic bladders. It has been used to treat high bladder pressures and incontinence due to both uninhibited contractions and sphincter dysfunction.

The first reported use of BoNT-A was in 1997 by Steinhardt *et al.* [77]. They injected 20U of BoNT-A in a 7-year-old girl with recurrent urinary tract infections (UTI) and bed-wetting as well as dramatic urethral dilatation due to severe external sphincter dyssynergia (ESD). After 18 months, the child was infection free and dry [77].

Since then, many other groups have documented the success of botulinum toxin in treating children with refractory urinary symptoms. In an American Academy of Paediatrics abstract in 1997, Maizels *et al.* presented a series of 20 females with a so-called "lazy bladder" and external sphincter spasticity who were treated with BoNT-A injections to the external sphincter [78]. At the 2-year follow-up, patients had no recurrent UTIs compared with a pre-procedural average of 3 infections per year. Maizel used 50U of BoNT-A and concomitant urethral dilatation. His results showed "subjective" improvements in daytime/nighttime flow and from this, he postulated that intravesical pressures post BoNT-A were likely decreased. Although the number of UTIs was reduced, Maizels did not achieve total continence in the patients in this study [78].

Radojicic *et al*, in 2006, treated 8 boys and 12 girls with recurrent UTIs, high post-void residuals (PVR), and fractional voiding with 50-100U of BoNT-A [79]. Follow-up was between 9 and 18 months, and all children were free of UTIs. One girl had transient urinary retention [79]. Mokhless *et al*, in 2006, prospectively evaluated 10 children with non-neurogenic neurogenic bladder [80]. All patients, except for one, were able to void without catheterization immediately after the procedure. Postoperative PVR decreased by 89%, and detrusor leak point pressures and uroflowometry were also considerably improved [80].

More recently, Franco *et al.* reviewed retrospectively 16 children with dysfunctional voiding who had treatment with BoNT-A [81]. They found that 75% of the children were dry at the first postoperative visit and 81% at the second. They also used a significantly higher dose (200-300IU) with increased efficacy and no increased morbidity. They concluded that endoscopic BoNT-A injection of the external sphincter appears to be a safe and effective way to treat refractory neurogenic voiding dysfunction in children with ESD [81].

Doses ranging from 50 to 300U of BoNT-A have been documented in the paediatric literature. Initially low doses were used to avoid toxicity and side effects. In the paediatric literature, the most serious side effects included a 10-day episode of urinary retention in one girl, a 2-day episode in another and 2 weeks of flank pain in one boy. All resolved spontaneously without the need for other treatment.

As the safety and efficacy of BoNT-A continues to be proven, paediatric urologists are beginning to use it as a treatment in children with refractory voiding dysfunction, daytime incontinence and non-neurogenic internal and external sphincter dyssynergia. Other treatment options that have been previously used in the adult population, including combined sphincteric and trigonal injection, are currently being evaluated in the paediatric population [82].

Administration of Botulinum Toxin

A number of variables are involved in the administration of BTx. These include the type and amount of BTx, the concentration, the number, depth and site of injections and even the instrumentation and type of anaesthetic used. Different units use different protocols, however, at the moment, evidence is lacking as to how best to administer BTx. With so many different variables to consider, establishing an "optimal protocol" is challenging!

Type of Botulinum Toxin

Most of the work on detrusor overactivity has used BoNT-A, with the BOTOX® formulation (Allergan Inc.) having been studied the most. There are a smaller number of published studies that have used Dysport (Ipsen, France). At the time of writing, BoNT-A is exclusively used in clinical practice within the field of Urogynaecology.

In comparison, only a few studies have looked at the efficacy of Botulinum toxin type B (BoNT-B). In 2003, Dyskra *et al.* set out to determine the efficacy and safety of BoNT-B (Myobloc) in 15 women with overactive bladder symptoms [83]. Five different doses of BoNT-B were administered: 2500U (5/15), 3750 (4/15), 5000 (2/15), 10,000U (2/15) and 15,000 (3/15). Clinical response was defined as subjective improvement in urinary frequency, urgency and incontinence symptoms [83]. 93% (14/15) of women showed clinical improvement. The degree of response was similar across all doses; however, the duration of the response was dose dependent. No serious side effects were reported, however, mild general malaise and dry mouth were seen in 2 of the 3 patients receiving the 15,000U dose. The group concluded that BoNT-B shows short-term efficacy; however, further double blind, placebo-controlled trials are warranted, before conclusions can be drawn.

Two further groups have also studied the efficacy of BoNT-B [84] [85]. Hirst *et al.* looked at the efficacy and safety of 5000U BoNT-B in 25 patients; 20 with urodynamic proven IDO and 5 with NDO [84]. Only 7 patients (28%), all with IDO, reported symptomatic improvement. This improvement was short-term, as symptoms started to return at the 10-week follow-up. They concluded that BoNT-B has a limited duration of action, and as a result, is unlikely to gain widespread use in the treatment of DO [84].

Ghei *et al.* tested the efficacy and safety of BoNT-B for the treatment of overactive bladder in a randomized, double blind, placebo-controlled, crossover trial [85]. Twenty patients with refractory DO were injected with either placebo or 5000U BoNT-B. After 6 weeks, the treatments were crossed over. Statistically significant paired differences in the change of average voided volume, urinary frequency and episodes of incontinence between active treatment and placebo were noted. They concluded that this study provides evidence for the efficacy of BoNT-B, however, again commented that its short duration of action is likely to limit its use [85]. One concerning fact was that 4 patients reported autonomic side effects, suggesting clinically relevant systemic spread of the toxin.

From these few studies, it appears that although BoNT-B has some clinical efficacy in the treatment of overactive bladder symptoms, its duration of action and possible systemic spread is a concern. Further larger randomised studies are needed, in particular, to directly compare BoNT-A and BoNT-B. Until then, BoNT-B remains in the domain of clinical trials, and not for use in clinical practice.

Concentration and Injected Volume

The concentration of BoNT-A injected depends, to a certain degree, on the condition being treated. Patients receiving treatment for NDO tend to receive higher concentrations than those with refractory IDO. Different units also have different protocols and even clinical studies vary in the concentration of BoNT-A used. To date, there is still not enough consistent evidence from clinical trials to provide a set protocol for the optimum dose of BoNT-A. The same applies to the amount of fluid injected into the bladder.

In 2008, a European expert panel consensus conference was convened with the main aim of evaluating the evidence and clinical considerations for the use of BoNT-A in the treatment of urologic and pelvic-floor disorders and to propose relevant recommendations. The published consensus report provides a summary of all studies performed up until 2008, and probably provides the best evidence to date for the optimum concentration of BoNT-A to use in each disorder, the dilution of toxin and the amount of liquid injected into the bladder [86].

The summarized results showed that 300U of BOTOX® was the most commonly used dosage in the treatment of NDO (1018 patients studied). The majority of these patients had either SCI or MS. Where Dysport was used, a lack of consensus remained as to the optimum concentration, with varying doses of 500, 750 and 1000U being used [86]. Up until 2008, a total of 641 patients with refractory IDO have received BoNT-A. The majority of these patients received 200U of BOTOX® into the detrusor. In the 2 studies where Dysport was used, 500U was injected [86].

Only 71 patients received treatment for painful bladder syndrome/Interstitial cystitis, reflecting the lack of published data. Each study used 100-200U of Botox®, but with varying dilutions [86]. Where BoNT-A was injected into the external urethral sphincter for the treatment of detrusor sphincter dyssynergia (640 patients), smaller concentrations were used. Most studies used Botox®, with doses ranging between 50U and 200U, however, only one study actually used 200U. The standard dose appeared to be 100U reconstituted in 4mls of normal saline [86]. Currently, there is no consensus as to the optimum dose for the treatment of pelvic floor disorders such as pelvic muscle spasm or chronic perineal pain.

Few studies have directly compared different doses of BoNT-A in terms of efficacy and side effects. Schurch *et al.* randomised 59 patients with NDO to receive either 200U or 300U of BoNT-A. No difference in primary outcomes was demonstrated, however, the authors were cautious in claiming that 200U is just as effective as 300U as study numbers were small [13].

Kuo *et al.* reported a randomised comparison of 100U, 150U and 200U in the treatment of 75 patients with refractory DO [87]. Clinical and urodynamic outcomes were similar between the 150U and 200U groups, with those patients receiving 100U experiencing less favorable therapeutic results. They concluded that 150U is just as effective as 200U but with less adverse events [87].

The amount of fluid injected at each site has also come under question, as again there is no common protocol. Initial studies diluted 100U of Botox® in 10mls of normal saline, a concentration of 10 U/ml [55]. Although more concentrated dilutions have been used, most experts agree that a higher volume provides better coverage of the bladder [88]. Based on the current literature, 10 to 30mls of total volume injected appears to be the standard treatment, with higher doses typically using the higher end of this range [88]. The European consensus report recommends that the dilution of Botox should be 10U/ml per site [86]. The optimum dose for the dilution of Dysport has yet to be determined.

Despite increasing evidence from reported literature, further collaborative and larger placebo-controlled trials are needed to define optimum dosing and injection protocols. Further direct comparison studies, looking at the efficacy of different doses of BoNT-A, will allow for the optimum treatment of patients, based on best clinical evidence, rather than on guesswork.

Number of Injections

Aside from study variations and individual preferences, the number of injection sites used will also depend on the condition, dilution and dose of BoNT-A used. No direct comparative studies exist as to the optimum number of injection sites per dose/dilution. Based on the current published literature, most studies use 20-30 injection sites.

One group has reviewed the existing literature pertaining to NDO and found that 61% (11 out of 18 studies) used 300U of Botox® injected over 30 sites (range 15-40) throughout the bladder [89]. No such comprehensive review exists for published studies on IDO, however, observational evidence suggests that fewer injection sites are usually used as a result of the lower concentrations of Botox® given.

Fewer injections may result in the same efficacy but have a less painful and faster injection procedure. This is obviously of benefit to both clinicians and patients, especially if patients are undergoing this procedure without general anaesthetic. The long-term risk of fibrosis may also be reduced, however, the risk of a false puncture may be increased. Further, larger direct comparison studies are needed to answer these questions.

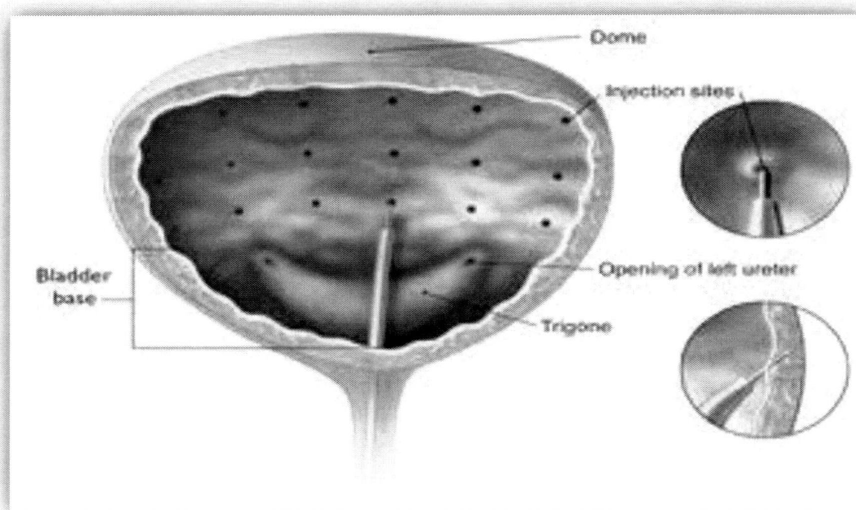

Figure 6. Example of an injection template showing injections throughout the posterior and lateral walls of the bladder, sparing the trigone and dome. Insets show an injection into the detrusor.

Site of Injections

The majority of studies and clinical practice involves the injection of BoNT-A into the lateral and posterior wall of the bladder. Initial studies advocated avoiding the dome and the trigone. The dome is the thinnest part of the bladder and so many suggest avoiding this area due to the risk of intraperitoneal injection and subsequent possible bowel injury [90].

One of the ongoing questions involves whether or not to inject the trigone. The initial decision to avoid the trigone was multifactorial. The trigone is innervated by a dense network of sensory, adrenergic and non-cholinergic pathways. It was thought that injection of this area might complicate the efficacy analysis of a cholinergic blockade. Subsequent basic and clinical research has suggested that sensory neuron dysfunction may actually contribute to the pathophysiology of DO and bladder pain syndrome [27] [29]. Since BoNT-A has been shown to inhibit both sensory neurons and the release of sensory neuropeptides [24] [27] [44], it seems logical that it may be of benefit to include the trigone in the injection distribution.

Studies have now started to investigate BoNT-A injection of the trigone. This year, a pilot study was performed at the University of Chicago to assess the subjective benefit of trigonal inclusion during BoNT-A injection [91]. Fourty patients with refractory NDO underwent trigone or trigone-sparing injection of 300U of Botox®. No difference between the 2 groups was noted, both in terms of clinical efficacy and side effects [91].

Taha *et al.* investigated 36 patients with refractory NDO due to SCI [92]. Patients were randomised into receiving either 300U of BoNT-A into the detrusor, sparing the trigone, or 200U of BoNT-A into the detrusor and 100U of BoNT-A into the trigone. Both arms showed improvement in all clinical parameters measured, however, these improvements were greatest in the combined arm: decrease in incontinence 80.9% vs 52.4%, complete dryness achieved in 66.7% vs 33.3% and decrease in quality of life score by 48.13% vs 46.76%. No patient showed new or upgraded vesicoureteric reflux or reported significant adverse events. The authors concluded from this that the superiority of including, rather than excluding, the trigone was significant [92].

Kuo compared the efficacy and safety of 100U of Onabotulinumtoxin A injection into the bladder body (37 patients), bladder body/trigone (35 patients) and bladder base/trigone (33 patients) [93]. Successful results were reported in 70%, 74% and 73% respectively. No significant differences in clinical effectiveness were noted between groups and the incidence of adverse events was similar. In particular, no vesicoureteric reflux was noted. Kuo concluded that bladder base/trigone injection is as safe and effective as bladder body injection with or without trigone injection [93].

Finally, as previously described, BoNT-A has been shown to have an antinociceptive effect on bladder afferent pathways. Pinto *et al.* studied the efficacy and tolerability of trigonal injection of BoNT-A in patients with painful bladder syndrome [94]. Urine concentration of NGF and brain-derived neurotrophic factor (BDNF) were also evaluated. Twenty-six women with refractory BPS/IC received 100U of BoNT-A, injected in 10 trigonal sites, a dose of 10U/ml. All patients reported subjective improvement at 1 and 3 month follow-up and treatment remained effective in >50% of the patients for 9 months. A significant, transient reduction in urinary NGF and BDNF was also observed. No cases of voiding dysfunction or vesicoureteric reflux occurred [94].

Evidence from recent studies suggests injection into the trigone is safe and effective. To date, no studies have reported evidence of voiding dysfunction or vesicoureteric reflux. However, the numbers enrolled in these studies are small and, therefore, information from larger, randomised controlled trials is needed before one can say conclusively that trigonal injection is safe.

Instrumentation and Anaesthesia

BoNT-A can be delivered with either a rigid or flexible cystoscope, depending on both patient and surgeon preference. Rigid systems allow for quicker, more controllable injections, whereas flexible scopes are generally more comfortable for patients. Flexible cystoscopy under local anaesthetic is ideal for the office setting and has a certain appeal for patients requiring repeated treatments. General Anaesthetic is more appropriate for anxious and sensitive patients and is exclusively required for most paediatric cases.

Repeated Treatments

Treatment with Botulinum toxin is not permanent. Significant improvements in symptoms are noted even at 2 weeks [60] [13] and, on average, last between 6 and 9 months. Again, the most comprehensive analysis of the duration of efficacy comes from the European Consensus report [86]. Certainly the condition being treated will have an impact on the duration of efficacy, however, it is thought that other variables such as the dose, concentration, number and site of injections may also play a role.

In those 1018 patients receiving BoNT-A for NDO, the mean duration of efficacy in single-injection studies was 8 months (range: 12-36 weeks) [86]. In comparison, the mean duration of efficacy was only 6 months (range: 4-10 months) in those patients being treated for IDO. There is limited evidence available from studies looking at the treatment for bladder pain syndrome, however, the best evidence available states that the duration of efficacy is probably less than 5months. In terms of injections into the external urethral sphincter, the duration of efficacy is the lowest, with clinical benefit being maintained for only 4.6 months [86].

Most patients will return for repeated injections once clinical symptoms return. There is limited information in the literature regarding the safety and continued efficacy of repeat injections of BoNT-A. Where studies have looked at repeated injections, it has only been in those patients receiving BoNT-A for NDO [95] [15] [96] [97]. However, the results are promising and suggest that clinical and urodynamic parameters remain constant after as many as 5 injections of Botox® [96] and 7 of Dysport [97]. Based on this evidence, it seems that repeated injections of BoNT-A are safe and may provide a valuable treatment option for patients with NDO.

There are reports of repeated injections causing immune responses in a small number of patients. Patients undergoing repeat injections are at risk of forming neutralizing antibodies that can interfere with the efficacy of BoNT-A therapy [98]. In time, these patients could become resistant to the effects of BoNT-A. As a result, studies are looking into the use of other BTx serotypes that do not cause immune reactions.

It should also be remembered that the long-term effects of BoNT-A are not known. Repeated injections could cause muscle fibrosis after several years and some studies have shown morphological and functional changes in muscles after BoNT-A injections [99]. There is a theoretical risk that patients may be left in a situation where they no longer respond to BoNT-A, due to antibody formation, however, have an even more unstable bladder due to pathological reinnervation. So although BoNT-A has excellent short-term effects, it should still be used with caution.

Side Effects

All drugs have side effects, and BoNT-A is no different. However, what is interesting, is that patients' acceptance of receiving BoNT-A injections is directly linked to the degree of side effects expected [2].

Significant side effects associated with BoNT-A injections are uncommon. Adverse events related to intradetrusor injection can be divided into local and systemic events.

Local Side Effects

Haematuria and post-operative pain are the most common symptoms observed following toxin injection. These complications are most likely due to the procedure itself, rather than the toxin. UTIs are also relatively common, occurring in up to 35% of patients [86], and again are likely due to the procedure.

BoNT-A injections can also cause extensive detrusor relaxation, resulting in detrusor underactivity and subsequent transient elevation of post-void residual volume and/or urinary retention [88]. Urinary retention is, understandably, of significant concern to patients. It is this risk, and the associated risk of CISC, that underlies the refusal of many patients to accept BoNT-A injections [2]. Reported rates are extremely variable, and will depend not only on the BoNT-A injection protocol used, but also on the study definition of "urinary retention" [88]. The rates for CISC use are also extremely variable, with studies quoting between 4 and 45% [86]. Again, study design and the concentration of BoNT-A used will influence the result.

This variability makes it very difficult to fully counsel patients about the risks. A 4% risk of CISC is completely different from a 45% risk! There are extensive studies that can be used to quote side effects, however, the most meaningful figures come from those review articles that have summarized the evidence from numerous published studies. Nitti quotes a 5% overall risk of "urinary retention" in patients with IDO and a subsequent16% risk of patients requiring CISC [88]. Popat *et al.* report de novo CISC in 69% of patients with NDO receiving 300U BoNT-A and 19% of IDO patients receiving 200U BoNT-A [100]. Kuo *et al.* showed a dose-dependent increase in voiding difficulty and acute retention: 0% with 100U, 10% with 150U and 20% with 200U [87].

Unless we work in a specialized spinal unit, the majority of our patients will be those with refractory IDO. It seems appropriate to quote a CISC risk in the region of 20% (1 in 5) based on the evidence from the literature, and indeed that is the risk quoted in our department. This risk is probably higher in patients with NDO due to spinal cord injury. As this risk is significant, women should be taught to perform CISC prior to receiving BoNT-A. For some, performing CISC will not be possible (eg severe arthritis hampering dexterity, dementia etc.), and, therefore, these patients should receive alternative treatment.

Systemic Side Effects

Systemic side effects have been reported with other uses of BoNT-A, however, very few have been reported in urology patients. De Laet *et al.* reviewed and summarized the published data on botulinum toxin and concluded that systemic side effects after BoNT-A injections for voiding disorders are very rare, but if they occur, can be disabling for patients, especially those with SCI [101]. Paraplegic patients can become completely dependent when they get upper extremity weakness and tetraplegic patients with high spinal cord lesions may develop difficulties with breathing.

Some case reports have described the paralysis of distal muscle groups after injection of BoNT-A into the detrusor or external sphincter [101]. Del Popolo reported muscle weakness in 5 patients receiving 300U BoNT-A [97] and Wyndaele reported upper limb weakness in 2 patients, also following 300U BoNT-A [102]. In total, 2.2-6% of cases reported some degree of muscle weakness. Other reported systemic side effects include flu-like symptoms, dry mouth (10%) and general malaise (5%) [101].

Serious side effects include Botulism and allergic reactions. Symptoms associated with Botulism are rare, and have been discussed at the beginning of this chapter. Symptoms of an allergic reaction can range from an unexplained rash or itching, to swollen airways and difficulty breathing. Thankfully this is also rare. To date, there is no report of Botulism after intradetrusor injection of BoNT-A in the published literature.

Contraindications to BoNT-A Administration

Certain groups of patients should not be given BoNT-A and it certainly should be used with extreme caution in others. There are no trials that look at the safety of BoNT-A in pregnancy and breastfeeding, and, therefore, BoNT-A should be avoided in these women.

BoNT-A should not be given to patients with pre-existing neuromuscular disease, such as myasthenia gravis, Eaton – Lambert syndrome and amyotrophic lateral sclerosis. Patients should also be informed that some formulations of BoNT-A include stabilizers, such as Albumin, derived from human blood, which may be of religious or cultural significance.

Patients' Views and Expectations of Botulinum Toxin

Patients' expectations are often neglected in clinical practice. Currently, studies are still looking at identifying the most appropriate patients to treat, the optimum dose, administration technique, frequency of treatment, as well as long term safety and efficacy. Until larger studies and long-term data are available, it is important that clinicians accurately select patients to receive BTx and fully discuss, inform and counsel them about the treatment.

Patient expectations, as well as treatment perspectives, have been demonstrated not to match with those of clinicians in the treatment of lower urinary tract disorders. A study by Digesu et al. in 2011, set out to explore patients' expectations about the outcomes and complications of botulinum toxin treatment for overactive bladder symptoms [2].

261 women with OAB and IDO were studied. Group A, 224 women, were treatment naïve and group B, 37 women, were non-responders to anticholinergics. Only 49.6% of women in group A and 54% of women in group B would accept BTx if offered to them [2].

In the treatment naïve group, up to 75% of women would accept BTx treatment only if the risk of voiding difficulties and using clean intermittent self-catheterization (CISC) was ≤ 15%. 78% of women would accept using CISC only for 1 day and 76% would accept and expect a beneficial effect lasting ≥ 9 months. Almost 90% of women would consider a cure rate ≥ 70% acceptable [2].

The reasons why these women declined to undergo botulinum toxin were as follows:

1) "Fear of being unable to empty my bladder normally": 10%
2) "Risk of needing to use a catheter to empty my bladder": 18%
3) "Beneficial effect that lasts less than 1 year": 8%
4) "Fear of undergoing any kind of anaesthesia": 6%

In group B, 76% of women would accept a risk ≤ 15% of being unable to empty their bladder and using CISC, 57% would accept using CISC no more than 1 day and up to 76% would consider it acceptable to have a beneficial effect ≥ 9 months. 86% of these women considered an acceptable cure rate ≥ 70%. Reasons for not accepting BTx treatment were similar to group A, with no significant difference noted [2].

This survey showed that the main reason women declined BTx treatment was the risk of voiding difficulties and needing to use CISC [2]. This highlights the need for clinicians to fully counsel patients to avoid false expectations. Although BTx is becoming a promising treatment for urinary tract symptoms, women are reticent to use it due to the current reported efficacy and side effects. Before BTx treatment is accepted as a widespread treatment modality, caution should be exercised until good-quality evidence from large-scale randomised controlled trials are reported.

Future Work

It is clear from the evidence in the literature that BoNT-A is a safe and effective treatment for a number of urogynaecological conditions. There is still work to be done in terms of identifying the optimum dose, concentration, site and number of injections. This will allow the development of a standardized protocol that can be used across all centers. This in turn will allow for direct comparisons of treatment outcomes and side effects between centers. One of the difficulties at present is that research groups use different doses and injection protocols, which make direct comparisons difficult.

At present, the overwhelming majority of botulinum toxin used within Urogynaecology is serotype A, with a few reported uses of serotype B. With the concern surrounding the development of antibodies with repeat treatments, it seems sensible for research to continue into whether other serotypes are equally as safe and effective.

BoNT-A will continue to be used "off-label" for IDO and pelvic floor muscle spasm until the FDA grants approval for its use. With continued reports of good clinical outcomes and minimal side effects, this is sure to come in the future.

References

[1] Verhayden J, Blitzer A and Brin MF (2002). Other non-cosmetic uses of Botox. *Semin Cutan Med Surg* 20: 121-126.

[2] Digesu GA, Panayi D, Hendricken C, Camarata M, Fernando R and Khullar V (2011). Women's perspective of botulinum toxin treatment for overactive bladder symptoms. *Int Urogynecol J* 22: 425-431.

[3] Erbguth FJ (2004). Historical notes on botulism, Clostridium botulinum, botulinum toxin and the idea of the therapeutic use of the toxin. *Movement disorders* 19(58): 52-56.

[4] Wordpress (2012). History of Botox. www.historyofbotox.com.

[5] Wikipedia (2012). *Botulinum toxin.* www.wikipedia.org/wiki/ Botulinum_toxin.

[6] Burgen A, Disckens F and Zatman L (1949). The action of Botulinum toxin on the neuro-muscular junction. *J Physiol* 109: 10-24.

[7] Clark RP and Berris CE (1989). Botulinum Toxin: A treatment for facial asymmetry caused by facial nerve paralysis. *Plastic and Reconstructive surgery* 84: 353-355.

[8] Carruthers JD and Carruthers JA (1992). Treatment of glabellar frown lines with C.Botulinum –A – exotoxin. *The Journal of Dermatologic Surgery and Oncology* 18: 17-21.

[9] Scott AB et al. (1994). Botulinum toxin for achalasia. Lancet 341: 244-245.

[10] Bushara KO and Park DM (1994). Botulinum toxin and sweating. *Journal of Neurology, Neurosurgery and Psychiatry* 57: 1437-1438.

[11] Dykstra DD, Sidi AA, Scott AB, Pagel JM and Goldish GD (1988). Effects of Botulinum A toxin on detrusor sphincter dyssynergia in spinal cord injury patients. *J Urol* 139: 919-922.

[12] Schurch B, Schmid DM and Stoher M (2000). Treatment of neurogenic incontinence with Botulinum toxin A. *N Engl J Med* 342: 665.

[13] Schurch B et al. (2005). Botulinum toxin type A is a safe and effective treatment for neurogenic urinary incontinence: results of a single treatment, randomized, placebo-controlled 6 month study. *J Urol* 174: 196-200.

[14] Reitz A, Stoher M, Kramer G, Popolo GD et al. (2004). European experience of 200 cases treated with botulinum A toxin injections into the detrusor muscle for urinary incontinence due to neurogenic detrusor overactivity. *Eur Urol* 45: 501-505.

[15] Karsenty G, Reitz A, Lindermann G, Boy S and Schurch B (2006). Persistence of therapeutic effect after repeated injections of botulinum toxin yype A to treat incontinence due to neurogenic detrusor overactivity, *Urology* 68: 1193-1197.

[16] Lacy DB, Tepp W, Cohen AC, DasGupta BR and Stevens RC (1998). Crystal structure of Botulinum Toxin type A and implications for toxicity. *Nat Struct Biol* 5(10): 898-902.

[17] Schulte-Boukloh H and Knispel HH (2006). Neuromodulatory therapies in female pelvic medicine and reconstructive surgery: biological agents. *BJU International* 98(51): 50-60.

[18] Rowland LP (2002). Stroke, spasticity and botulinum toxin. *N Engl J Med* 347(6): 382-383.

[19] Sacco E, Pinto F and Bassi P (2008). Emerging pharmacological targets in overactive bladder therapy: experimental and clinical evidences. *Int Urogynecol J* 19:583-598.

[20] Gur S, Kadowitz PJ and Hellstrom WJG (2007). Purinergic (P2) receptor control of lower genitourinary tract function and new avenues for drug action: an overview. *Curr Pharm Design* 13: 3236-3244.

[21] Yu Y and DeGroat WC (2008). Sensitization of pelvic afferent nerves in the in vitro rat urinary bladder-pelvic nerve preparation by purinergic agonists and cyclophosphamide pretreatment. *Am J Physiol Renal Physiol.*

[22] Kumar V, Chapple CR et al. (2007). Enhanced adenosine triphosphate release from the urothelium of patients with painful bladder syndrome: a possible pathophysiological explanation. *J Urol* 178: 1533-1536.

[23] Khera M, Somogyi GT, Kiss S, Boone TB and Smith CP (2004). *Botulinum toxin A inhibits ATP release from bladder urothelium after chronic spinal cord injury* 45: 987-993.

[24] Apostolidis A, Dasgupta P and Folwer C (2006). Proposed mechanism for the efficacy of injected botulinum toxin in the treatment of human detrusor overactivity. *Eur Urol* 49: 644-650.

[25] Atiemo H, Wynes J, Chuo J et al. (2005). Effect of Botulinum toxin on detrusor overactivity induced by intravesical adenosine triphosphate and capsaicin in a rat model. *Urology* 65: 622-626.

[26] Chuang YC, Yoshimura N, Huang CC et al. (2004). Intravesical Botulinum toxin A administration produces analgesia against acetic acid induce bladder pain responses in rats. *J Urol* 172: 1529-1532.

[27] Apostolidis A, Popat R, Yiangou Y et al. (2005). Decreased sensory receptors P2X3 and TRPV1 in suburothelial nerve fibres following intradetrusor injections of Botulinum toxin for human detrusor overactivity. *J Urol* 174: 977-983.

[28] Blok BF (2002). Central pathways controlling micturition and urinary incontinence. *Urology* 59: 13-17.

[29] Fowler CJ (2002). Bladder afferents and their role in the overactive bladder. *Urology* 59: 37-42.

[30] Cruz F, Guimaraes M, Silva C et al. (1997). Desensitisation of bladder sensory fibres by intravesical capsaicin has long lasting clinical and urodynamic effects in patients with hyperactive or hypersensitive bladder dysfunction. *J Urol* 157: 585-589.

[31] Silva C, Ribeiro MJ and Cruz F (2002). The effect of intravesical resiniferatoxin in patients with idiopathic detrusor instability suggests that involuntary detrusor contractions are triggered by C-fiber input. *J Urol* 168: 575-579.

[32] Cruz F (2004). Mechanisms involved in new therapies for overactive bladder. *Urology* 63: 65-73.

[33] Brady CM, Apostolidis A, Harper M et al. (2004). Parallel changes in bladder suburothelial vanilloid receptor TRPV1 and pan-neuronal marker PGP9.5 immunoreactivity in patients with neurogenic detrusor overactivity after resiniferatoxin treatment. *BJU Int* 93: 770-776.

[34] Brady CM, Apostolidis A, Yiangou Y et al. (2004). P2X3 immunoreactive nerve fibres in neurogenic detrusor overactivity and the effect of intravesical resiniferatoxin. *Eur Urol* 46: 247-253.

[35] Silva C, Silva J, Ribeiro MJ et al. (2005). Urodynamic effect of intravesical resiniferatoxin in patients with neurogenic detrusor overactivity of spinal origin: results from a double-blind, randomized, placebo-controlled trial. *Eur Urol* 48: 650-655.

[36] Silva C, Silva J, Castro H et al. (2007). Bladder sensory desensitization decreases urinary urgency. *BMC Urol* 7: 9.

[37] Morenilla-Palao C et al. (2004). Regulated exocytosis contributes to protein kinase C potentiation of vanilloid receptor activity. *J Biol Chem* 279: 25665-25672.

[38] Ito K, Rome C, Bouleau Y and Dulon D (2002). Substance P mobilizes intracellular calcium and activates a nonselective cation conductance in rat spinal ganglion neurons. *Eur J Neurosci* 16: 2095-2102.

[39] Paukert M, Osteroth R, Geisler HS et al. (2001). Inflammatory mediators potentiate ATP-gated channels through the P2X3 subunit. *J Biol Chem* 276: 21077-21082.

[40] LucioniA, Bales GT, Lotan TL et al. (2008). Botulinum toxin type A inhibits sensory neuropeptide release in rat bladder models of acute injury and chronic inflammation. *BJU Int* 101: 366-370.

[41] Welch MJ, Purkiss JR and Foster KA (2000). Sensitivity of embryonic rat dorsal ganglia neurons to Clostridium botulinum neurotoxins. *Toxicon* 38: 245-258.

[42] Lowe EM, Anand P, Terenghi G et al. (1997). Increased nerve growth factor levels in the urinary bladder of women with idiopathic sensory urgency and interstitial cystitis. *B J Urol* 79: 572-577.

[43] Yoshimura N (1999). Bladder afferent pathway and spinal cord injury: possible mechanisms inducing hyperreflexia of the urinary bladder. *Prog Neurobiol* 57: 583-606.

[44] Giannantoni A, Di Stasi SM, Nardicchi V et al. (2006). Botulinum A toxin injection into the detrusor decrease berve growth factor bladder tissue levels in patients with neurogenic detrusor overactivity. *J Urol* 175: 2341-2344.

[45] Lowe EM, Anand P, Terenghi G et al. (1997). Increased nerve growth factor levels in the urinary bladder of women with idiopathic sensory urgency and interstitial cystitis. *B J Urol* 79: 572-577.

[46] Liu HT, Chancellor MB and Kuo HC (2009). Urinary growth factor levels are elevated in patients with detrusor overactivity and decreased in responders to detrusor botulinum toxin-A injection. *Eur Urol* 56: 700-707.

[47] Abrams P et al. (2005). *Recommendations of the International Scientific committee: Evaluation and treatment of Urinary incontinence, pelvic organ prolapse and faecal incontinence.*

[48] Gallien P, Robineau, S Verin M et al. (1998). Treatment of detrusor sphincter dyssynergia by transperineal injection of botulinum toxin. *Arch Phys Med Rehabil* 79: 715-717.

[49] Vapnek JM, Couillard DR and Stone AR (1994). Is sphincterotomy the best management of the spinal cord injured bladder? *J Urol* 151:961-964.

[50] Phelan MW, Franks M, Somoyogi GT et al. (2001). Botulinum toxin urethral sphincter injection to restore bladder emptying in men and women with voding dysfunction. *J Urol* 165: 1107-1110.

[51] Schurch B, Hauri D and Rodic B (1996). Botulinum-A toxin as a treatment od detrusor sphincter dyssynergia: a prospective study in 24 spinal cord injury patients. *J Urol* 155: 1023-1029.

[52] Wheller JS, Walter JR, Chintam RS et al. (1988). Botulinum toxin injections for voiding dysfunction following SCI. *J Spinal Cord Med* 21: 227-229.

[53] Abrams P, Cardozo L, Fall M et al. (2002). The standardization of terminology of lower urinary tract functions: report from the Standardisation Sub-committee of the International Continence Society. *Neurourol Urodyn* 21: 167-178.

[54] Irwin DE, Milsom I et al. (2006). Population-based survey of urinary incontinence, overactive bladder and other lower urinary tract symptoms in five countries: results of the EPIC study. *Eur Urol* 50: 1306-1314.

[55] Schurch B, Stohrer M and Kramer D (2000). Botulinum A toxin for treating detrusor hyperreflexia in spinal cord injured patients: a new alternative to anticholinergic drugs? Preliminary results. *J Urol* 164: 692-697.

[56] Giannantoni SM, Di Stasi RL et al. (2004). Intravesical resiniferatoxin versus Botulinum A toxin for neurogenic detrusor overactivity: a prospective randomized study. *J Urol* 172: 240-243.

[57] Schurch B et al. (2005). Botulinum toxin type A is a safe and effective treatment for neurogenic urinary incontinence: results of a single treatment, randomized, placebo-controlled 6 month study. *J Urol* 174: 196-200.

[58] Herschon S et al. (2011). Efficacy of botulinum toxin A injection for neurogenic detrusor overactivity and urinary incontinence: a randomized, double-blind trial. *J Urol* 185: 2229-2235.

[59] Apostolidis A et al. (2009). Recommendations on the use of Botulinum toxin in the treatment of lower urinary tract disorders and pelvic floor dysfunctions: A European Consensus Report. *Eur Urol* 55: 100-120.

[60] Kuo HC (2004). Urodynamic evidence of effectiveness of Botulinum A toxin injection in treatment of detrusor overactivity refractory to anticholinergic agents. *J Urol* 63(5): 868-872.

[61] Schmid DM, Sauermann P et al. (2006). Experience with 100 cases treated with botulinum A toxin injections in the detrusor muscle for idiopathic overactive bladder syndrome refractory to anticholinergics. *J Urol* 176: 177-185.

[62] Sahai A, Dowson C, Khan MS and DasGupta P (2009). Improvement in quality of life after botulium toxin-A injections for idiopathic detrusor overactivity: results from a randomized double-blind placebo-controlled trial. *BJU Int* 103: 1509-1515.

[63] Flynn MK, Amundsen CL et al. (2009). Outcome of a randomized, double-blind, placebo-controlled trial of botulinum A toxin for refractory overactive bladder. *J Urol* 181: 2608.

[64] Brubaker L, Richter HE et al. (2008). Refractory idiopathic urge urinary incontinence and botulinum A injection. *J Urol* 180: 217.

[65] Van de Merwe JP, Nordling J et al. (2008). Diagnostic criteria, classification and nomenclature for painful bladder syndrome/interstitial cystitis: an ESSIC proposal. *Eur Urol* 53(1): 60-67.

[66] Smith CP et al. (2004). Botulinum toxin A has antinociceptive effects in treating interstitial cystitis. *Urology* 64: 871-875.

[67] Kuo HC (2005). Preliminary results of suburothelial injection of botulinum A toxin in the treatment of chronic interstitial cystitis. *Urol Int* 75(2): 170-174.

[68] Giannantoni A, Costantini E et al. (2006). Botulinum A toxin intravesical injections in the treatment of painful bladder syndrome: a pilot study. *Eur Urol* 49: 704-709.
[69] Giannantoni A, Porena M, Costantini E et al. (2008). Botulinum A toxin intravesical injections in the treatment of painful bladder syndrome: 1 year follow-up. *J Urol* 179: 1031-1034.
[70] Thomson AJ, Jarvis SK, Lenart M et al. (2005). The use of botulinum type A (BOTOX) as treatment for intractable chronic pelvic pain associated with spasm of the levator ani muscles. *BJOG* 112: 247-249.
[71] Brown CS, Glazer HI, Vogt V et al. (2006). Subjective and objective outcomes of botulinum toxin type A treatment in vestibulodynia: pilot data. *J Reprod Med* 51(8) 635-641.
[72] Abbott JA, Jarvis SK, Lyons SD et al. (2006). Botulinum toxin type A for chronic pain and pelvic floor spasm in women: a randomized controlled trial. *Obstet Gynecol* 108: 915-923.
[73] Read S, King M and Watson J (1997). Sexual dysfunction in primary medical care: prevalence, characteristics and detection by the general practitioner. *J Public Health Med* 19: 387-391.
[74] Brin MR and Vapnek JM (1997). Treatment of Vaginismus with botulinum toxin injection. *The Lancet* 349:252-253.
[75] Shafik A and El-Sibai O (2000). Vaginismus: results of treatment with botulinum toxin. *J Obstet Gynaecol* 20:300-302.
[76] Ghazizadeh S and Nikzad M (2004). Botulinum toxin in the treatment of refractory Vaginismus. *Obstet Gynecol* 104: 922-925.
[77] Steinhardt GF, Naseer S and Cruz OA (1997). Botulinum toxin: novel treatment for dramatic urethral dilatation associated with dysfunctional voiding. *J Urol* 158: 190.
[78] Maizels (1997). *Abstract presented at the American Anatomy of paediatrics Annual Meeting.*
[79] Radojicic ZI, Perovic SU and Milic NM (2006). Is it reasonable to treat refractory voiding dysfunction in children with botulinum A toxin? *J Urol* 176(1): 332-336.
[80] Mokhless I, Gaafar S, Fouda K, Shafik M and Assem A (2006). Botulinum A toxin urethral sphincter injection in children with non-neurogenic neurogenic bladder. *J Urol* 176: 1767-1770.
[81] Franco I, Landau-Dyer L et al. (2007). The use of botulinum toxin A injection for the management of external sphincter dyssynergia in neurogenically normal children. *J Urol* 176: 1775-1780.
[82] Dyer LL and Franco I (2009). Botulinum toxin A therapy in paediatric Urology: Indications for the neurogenic and non-neurogenic neurogenic bladder. *TheScientificWorldJOURNAL: TSW Urology* 9: 1300-1305.
[83] Dyskra D, Enriquez A and Valley M (2003). Treatment of overactive bladder with botulinum toxin type B: a pilot study. *Int Urogynecol J* 14:424-426.
[84] Hist GR, Watkins AJ, Guerrero K et al. (2007). Botulinum toxin B is not an effective treatment of refractory overactive bladder. *Urology* 69: 69-73.

[85] Ghei M, Maraj BH, Miller R et al. (2005). Effects of botulinum toxin B on refractory detrusor overactivity: a randomised, double-blind, placebo-controlled crossover trial. *J Urol* 174: 1873-1877.
[86] Apostolidis A et al. (2009). Recommendations on the use of botulinum toxin in the treatment of lower urinary tract disorders and pelvic floor dysfunction: A European Consensus report. *Eur Urol* 55: 100-120.
[87] Kuo HC (2005). Comparison of the therapeutic effects of suburothelial injection of botulinum A toxin at different doses for refractory detrusor overactivity. *J Urol* 175:113.
[88] Nitti VW (2006). Botulinum toxin for the treatment of Idiopathic and Neurogenic overactivity. Bladder: state of the art. *Rev Urol* 8(4) 198-208.
[89] Karsenty G, Denys P et al. (2008). Botulinum toxin A (Botox) intradetrusor injections in adults with neurogenic detrusor overactivity/neurogenic overactive bladder. A systematic literature review. *Eur Urol* 53(2): 231-456.
[90] Rapp DE, Lucioni A and Bales GT (2007). Botulinum toxin injection: a review of injection principles and protocols. *Int Braz J Urol* 32(2).
[91] Lucioni A, Rapp DE, Gong EM et al. (2012). *Botulinum toxin type A injections for overactive bladder: trigone versus trigone-sparing injections.* In Press.
[92] Abdel-Meguid TA (2010). Botulinum toxin A injections into neurogenic overactive bladder – to include or exclude the trigone? A prospective, randomised controlled trial. *J Urol* 184(6) 2578-2585.
[93] Kuo HC (2011). Bladder base/trigone injection is safe and as effective as bladder body injections of Onabotulinumtoxin A for idiopathic detrusor overactivity refractory to antimuscarinics. *Neurourol Urodyn* 30(7) 1242-1248.
[94] Pinto R, Lopes T, Frias B et al. (2010). Trigonal injection of Botulinum toxin A in patients with refractory bladder pain syndrome/interstitial cystitis. *Eur Urol* 58: 360-365.
[95] Grosse J, Kramer G and Stohrer M (2005). Success of repeat detrusor injections of botulinum A toxin in patients with severe neurogenic detrusor overactivity and incontinence. *Eur Urol* 47: 653-659.
[96] Reitz A, Denys P, Fermanian C et al. (2007). Do repeat intradetrusor Botulinum toxin A injections yield valuable results? Clinical and urodynamic results after 5 injections in patients with neurogenic detrusor overactivity. *Eur Urol* 52(6): 1729-1735.
[97] Del Popolo G, Filocamo MT et al. (2008). Neurogenic detrusor overactivity treated with English botulinum toxin A: 8 year experience of one single centre. *Eur Urol* 53: 1013-1020.
[98] Moore C, Rackley R and Goldman H (2005). Urologic applications of Botox. *Current Urology reports* 6: 419-423.
[99] Kranjc BS, Sketelj J et al. (2001). Long term changes in myosin heavy chain composition after botulinum A toxin injection into rat medial rectus muscle. *Invest Opthalmol Vis Sci 42*: 1358-1364.
[100] Popat R, Apostolidis A et al. (2005). A comparison between the response of patients with IDO and NDO to the first intradetrusor injection of botulinum toxin A injection. *J Urol* 174: 984-989.

[101] De Leat K and Wyndaele JJ (2005). Adverse events after BONT-A toxin injection for neurogenic voiding disorders. *Spinal Cord* 43: 397-399.

[102] Wyndaele JJ and Van Dromme SA (2002). Muscular weakness as side effect of botulinum toxin injection for neurogenic detrusor overactivity. *Spinal Cord* 40: 599-600.

In: Botulinum Toxin
Editors: Fausto E. Marino and Sebastiano De Luca
ISBN: 978-1-62257-185-7

Chapter II

Effects of Botulinum Toxin on Neuronal Communication: Converging Evidences of Central Effects

C. Rasetti-Escargueil
National Institute for Biological Standards and Control
A Centre of Health Protection Agency
Blanche Lane, South Mimms, Potters Bar, UK

Abstract

Botulinum neurotoxins (BoNTs) are increasingly used thera-peutically to locally paralyse muscles for clinical or cosmetic benefit. Initially used to treat strabismus in the 1970s, botulinum toxin has more than a hundred possible medical applications: strabismus, movement disorders (dystonia and muscle hyperactivity), including hemifacial spasm, essential tremor, tics, cervical dystonia, cerebral palsy and vascular cerebral stroke. This deadly poison traditionally considered as a paralytic agent has soon been reported to provide pain relief prior to spasm relief. Despite recurrent clinical evidences and now approved use in chronic migraine there is only limited experimental evidence supporting botulinum toxin action on the central nervous system. Understanding the mechanism of action related to the use of BoNTs in the treatment of pain is still desired. This chapter will provide a more detailed understanding of current knowledge relating to the central actions of botulinum toxins, with particular emphasis on its potential therapeutic applications to pain management and human epilepsy.

Figure 1. Justinus Kerner (godfather of BoNT Research, *public domain from Wikimedia commons website*).

History of Botulinum Toxin: From Sausage Poisoning To Global Therapeutic and Cosmetic Use

The paralytic effects of botulinum toxin have been used since ancient times when shamans suggested to Indian maharajas a deadly formula for enemies:

> "Fill the gut of a sheep with the blood taken from the carotid artery of a black bull, tighten both ends, and hang it in the shadow of a mulberry tree. When dried grind the content firmly. A pinch of the powder mixed with food will kill the consumer within three days".

Because of delayed lethality, the host was not suspected. This sausage poisoning known later as "Kerner's disease" was first described in a manuscript published in 1822 where Kerner summarized 155 cases of "sausage poisoning" in Germany [1]. Kerner also envisioned the possibility of using this poison to treat diseases associated with an overactive nervous system, including muscle hyper-contraction and hyper-secretion of body fluids. The name of 'botulism' was coined in the second half of the 19th century from the latin word *botulus* meaning sausage by Van Ermengem at the University of Ghent. He isolated an anaerobic sporulating microorganism that he named Bacillus botulinus. Macerated ham samples, filtrates of macerated ham, as well as cultures and culture filtrates fed to mice, guinea pigs, and monkeys caused symptoms, comparable to those found in the human victims, and death. These observations were published in 1897 in a German microbiological journal [2]. The term Clostridium was then used to designate anaerobic spore-forming bacteria as opposed to Bacillus species which only includes aerobic microorganisms [3, 4].

Botulinum neurotoxins (BoNT) produced under anaerobic conditions by certain strains of *Clostridium botulinum*, *Clostridium baratii*, and *Clostridium butyricum*, are considered the most lethal biological toxic substances known to man. Of the seven known BoNTs, serotypes A, B, and E account for almost all human cases of botulism.

Table 1. Timeline for BoNT (up-dated from Hanchanale et al. 2010)

886 AD	Emperor Leo IV	banned 'blood sausage' as it caused fatal illness
1822	Justinus Kerner	sausage poison (possible therapeutic use)
1870	Müller	botulism (Latin: botulus) for sausage
1895	Van Ermengem	*Clostridium botulinum* (causative agent of botulism)
1919	G.S. Burke	minimum lethal dose: guinea pigs
1928	Herman Sommer	BoNT (purified) isolation
1946	Carl Lamanna Edward Schantz	LD50 test: neurotoxin activity BTA in in crystalline form
1949	Arnold Burgen	Neuromuscular transmission blockade
1950	Vernon Brooks	BTA: blockade of acetylcholine from motor nerve endings
1960s	Schantz/Scott	strabismus: monkeys
1980	Scott	strabismus: humans
1984	FDA	designates BTX for blepharospasm in adults
1986	FDA	designates BTX for cervical dystonia
1987	Drs. Jean and Alastair Carruthers	Cosmetic benefits of BTX found accidentally by ophthalmologists treating patients for involuntary blinking
1988	Allergan	Oculinum (BTA): clinical trials
1989	FDA Allergan	Oculinum (BTA): strabismus, blepharospasm and dystonia (~ 25-40 ng/100 mouse LD_{50}) Allergan buys Oculinum and renames as Botox
1993		SNAP-25: molecular target of botulinum toxin type A
1995	MHRA	approves Dysport® (~5ng/500 mouse LD50) for strabismus in UK
2000	FDA	approves BTX for cervical dystonia
2001	- CPMP	BOTOX®Cosmetic approval for cosmetic procedures in Canada and New Zealand. approves first type B toxin, NeuroBloc® for cervical dystonia
2002	FDA	approves BTX for cosmetic therapy (Australia, Switzerland, Taiwan and Singapore)
2003	AFSSAPS	approves BTX as VISTABEL® (France)
2004	FDA	FDA approves BTX for primary axillary hyperhidrosis (severe underarm sweating)
2006	MHRA	approves BTX as VISTABEL® for treatment of glabellar lines
2006	-	Xeomin® (first pure A toxin) licensed in Germany for blepharospasm and cervical dystonia in adults
2010	FDA	approves Botox to treat chronic migraine
2011	FDA	approves Botox to treat specific form of urinary incontinence

FDA: Food and Drug administration; MHRA: Medicines and Healthcare products Regulatory Agency; CPMP: Committee for Proprietary Medicinal Products; AFSSAPS: French Agency for the Safety of Health products.

BoNTs are synthesized as inactive 150 kDa single-chain proteins that are activated by proteolytic cleavage to form a disulfide-linked dimer comprising a 100 kDa heavy chain and a 50 kDa light chain [5]. The heavy chain is responsible for neuronal targeting, whereas the smaller, light chain acts as a highly specific zinc-dependent protease that cleaves one or more components of the neuronal membrane fusion complex [6]. As a result of this cleavage event, the neuronal exocytosis docking/fusion machinery becomes impaired leading to a life-threatening flaccid paralysis [7]. Because of their high toxicity the toxins are classified as potential biological weapons or “Category A” agents by the Centre for Disease Control and Prevention (CDC) [8].

Nevertheless, minute doses of this deadly poison are increasingly used therapeutically to locally paralyse muscles for clinical or cosmetic benefit. Initially used to treat strabismus, botulinum toxin has now more than a hundred possible medical applications including movement disorders, hemifacial spasm, essential tremor, tics, writer’s cramp, cervical dystonia, cerebral palsy and vascular cerebral stroke. Botulinum toxin has recently been approved for chronic pain, migraine headache, overactive bladder and inflammation [9, 10].

Although initially thought to inhibit acetylcholine release only at the neuromuscular junction, botulinum toxins are now recognized to inhibit acetylcholine release at autonomic cholinergic nerve terminals, as well as peripheral release of neurotransmitters involved in pain regulation. Thus, their use in neurology has been expanded to headache and other pain syndromes, as well as hyper-secretory disorders, migraines and hyperhidrosis (Table 1).

Moreover, BoNT-A has long heritage of use in various treatment settings with no major safety concerns leading to confidence in its safety and efficacy. Its application does not induce persistent histological changes in the nerve terminal or the muscle after long term treatment [11].

Complex Mode of Action Involving Elaborated Neuronal Targeting and Modulation of Vesicle Trafficking Activity

BoNTs were initially shown to induce prolonged muscle paralysis by blocking the release of neurotransmitters from peripheral cholinergic nerve endings [12, 13]. This involves a concentration step at complex polysialo-gangliosides of the plasma membrane and penetration in recycling synaptic vesicles after binding to a specific protein receptor derived from synaptic vesicles. The neurotoxins then shut down the synaptic vesicle cycle by specific cleavage of protein core components of the cellular membrane fusion machinery. The uptake of four out of seven known BoNTs into synaptic vesicles has been demonstrated to rely on binding to synaptotagmin or synaptic vesicle protein 2 [14].

The heavy chain (HC) channel (depicted in red in figure 2) prevents the aggregation of the light chain (LC) cargo (depicted in cyan) in the acidic endosomal lumen, maintaining the unfolded or partially folded conformation (cyan ribbon) during translocation and releasing the LC after it refolds at the neutral cytosolic pH.

The LC is illustrated as both a ribbon and a space-filling shape to highlight the unfolding/refolding dynamics of cargo during transit and the associated transient interactions between cargo and channel. The disulfide linkage between the HC and the LC is shown in green. [15].

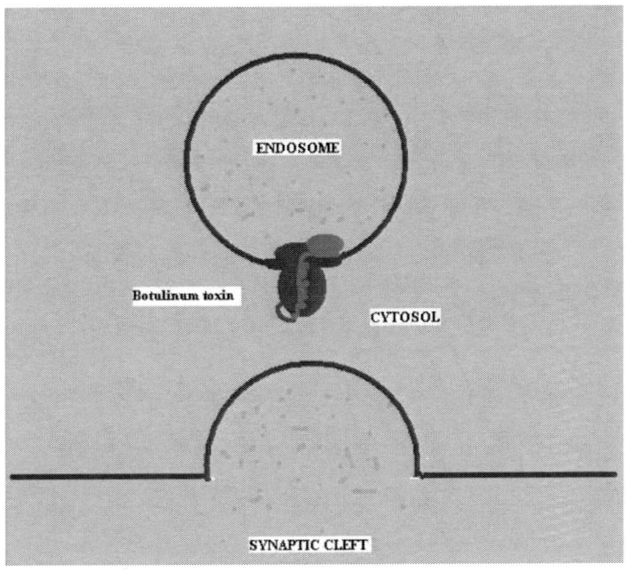

Figure 2. Sophisticated strategy to enter neurons: "*BoNT is an astonishing modular nanomachine that unites recognition, trafficking, unfolding, translocation, refolding, and catalysis in a single entity*."[15].

Actions of BoNT on peripheral nerve system have been extensively studied by Simpson and collaborators but surprisingly far more remains to be elucidated regarding BoNT action on the central nervous system. In addition, several modes of synaptic vesicle release and recycling have now been identified: 'full-collapse fusion', 'kiss-and-run' exocytosis or 'bulk endocytosis' [16] but new investigations are still needed to clarify botulinum toxin implications regarding different modes of synaptic release. Different BoNTs have slightly different modes of action [17], and different modes of transmitter release are differentially sensitive to BoNTs. It has even been reported that a late phase of exocytosis is not blocked by BoNTs [18]. The seven BoNTs all act by targeting the same vesicle fusion machinery but cause paralysis of different durations as a result of different intracellular behaviours [19]. BoNT serotypes B, D, F and G cleave specifically at different peptide bonds of the vesicle associated membrane protein (VAMP) synaptobrevin, a membrane protein of small synaptic vesicles (SSVs). BoNT types A, C and E cleave SNAP-25 at different sites located within the carboxyl-terminus, while BoNT type C additionally cleaves syntaxin. This remarkable specificity of BoNTs with differential distribution of synaptic vesicle protein receptors, gangliosides and toxin substrates in distinct neuronal populations opens up the possibility of using different serotypes of botulinum toxins for the treatment of various central nervous system diseases [20-22].

Evidence for Central Effects of Botulinum Toxin Type A

Historical distinction between botulinum and tetanus toxins established that BoNT remained localized to the injection site unlike tetanus toxin, but experimental work by Caleo's group has recently shown significant retrograde transport and transcytosis of BoNT across neurons in vivo [23-24].

As a consequence, a more detailed understanding of the central trafficking and actions of this neurotoxin is now required in particular for its potential therapeutic applications to human epilepsy and pain management [25-27].

Botulinum Toxin Action on Spinal Processing and Autonomic Nervous System

Despite the lack of evidence of direct action on central nervous system (CNS) after intramuscular injection in humans, several studies strongly suggest that BT-A affects the functional organization of the CNS indirectly through peripheral mechanisms. Muscle afferent input is tightly coupled to motor cortical output, so that the afferents from a stretched muscle go to cortical areas where they can excite neurons capable of contracting the same muscle. The BoNT-induced reduction in motor output could alter the balance between afferent input and motor output, thereby changing cortical excitability [28]. The toxin was also found to alter sensory inputs to the central nervous system, thus indirectly inducing secondary central changes. Some of the long-term clinical benefits of BoNT-A treatment may reflect plastic changes in motor output after the reorganization of synaptic density [29].

Neurophysiological changes after intramuscular injection of botulinum toxin have been recently described in detail by Palomar et al. [31] Clinical studies presented in this review demonstrate indirect action of the toxin on the excitability of spinal cord circuitry in humans, by acting on the pre-synaptic neuromuscular junction and by decreasing spindle afferent input to the spinal cord. The results also show that BoNT injected at therapeutic doses induces spinal and cortical effects through plastic rearrangement in the brain subsequent to denervation or alterations in sensory input. An excellent clinical example is the positron emission tomography (PET) study of the pattern of cortical activation induced by BT injections. PET activation studies show an over-activity of striatum and non primary motor areas with under-activity of the primary motor cortex in dystonia patients during voluntary movements. BoNT injections in those patients improved writing technique together with increased activation in the parietal cortex and caudal supplementary motor area, leaving the pattern of activity in the primary motor cortex unchanged. These results clearly show a cortical reorganization secondary to the BoNT induced changes in motor strategy [30-31].

Retrograde Transport and Transcytosis to Neurons in the Central Nervous System (CNS)

In vivo studies by Caleo and collegues have shown that active BoNT/A is retrogradely transported by central neurons and motoneurons and is even transcytosed to afferent synapses, in which it cleaves SNAP-25. SNAP-25 cleavage by BoNT/A was also observed in the contralateral hemisphere after unilateral BoNT/A delivery to the hippocampus resulting in blockade of hippocampal activity in the untreated hemisphere. By showing axonal migration and neuronal transcytosis of BoNT/A, these findings revealed a novel pathway of BoNT/A trafficking into neurons and opened new concepts in the clinical uses of this neurotoxin [32]. Using the visual pathway as a model system, the same group showed that catalytically active BoNT/A was capable of anterograde axonal transport following injection into the rat eye. Cleaved SNAP-25 was present in presynaptic structures of the tectum, but retinal terminals were devoid of the immunoreactivity, indicative of transcytosis. These data confirm the existence of unknown mechanisms of transport and action of BoNT/A [33].

Nevertheless, toxin doses used in experimental studies were extremely high compared to minute doses used in the clinic which supports the need for clinical research to determine if there is a central action of BoNT-A resulting from a peripheral administration [34]. To date, there is no clinical report showing that therapeutic doses of BoNT can cross the blood–brain barrier.

Effects of Botulinum Toxin on Pain

Initial Reports and Hypothesised Mechanism for the Antinociceptive Action

The first indication of beneficial effects of BoNT on pain emerged from patient's reports. BoNT was putting an end to the vicious circle of muscle spasm/pain, helping the return to normal activities with subsequent long-term recovery [35].

A proposed mechanism of action for pain relief was initially multifactorial including direct action on muscle and indirect effects via the neuromuscular junction.

The first hypotheses regarding how BoNT could directly or indirectly alter pain, included a changes in the sensitivity and response patterns of group III and IV muscle nociceptors, diminished activity in the gamma-motor neurons leading to changes in muscle spindle afferents, alterations in cholinergic control of vascular and autonomic functions, including neurogenic inflammation, neuroplastic changes in the processing of afferent somatosensory activity and direct non-cholinergic effects on pain afferents [36].

BoNT therapy is used in states of muscle hyperactivity such as contractures or spasm and focal dystonias. The "integrated hypothesis" suggests that a lesion of a muscle damages the neuromuscular endplate causing excessive acetylcholine release. The contracture compresses small blood vessels leading to tissue ischemia and subsequent release of bradykinin (BKN) which causes sensitization of nociceptors. Reflex spasm can be induced by nociceptive input from neighboring joints or muscles causing compression of the large blood vessels causing again ischemia. In cases of weak dystonias, a compression of blood vessels is unlikely. However, the tonic contraction will cause a lowering of pH and a release of ATP at concentrations sufficient to excite muscle nociceptors.

BoNT stops excessive acetylcholine release and in cases of spasm and dystonia, can abolish the pain by relaxing the muscle. Since many patients report alleviation of their pain before the muscle relaxing effect of BoNT, a direct analgesic action is not excluded suggesting that BoNT has a more complex mechanism of action than initially hypothesized.

Initial assumptions suggested that BoNT may also inhibit neurotransmitters release such as neuropeptides from nociceptive nerve.

BoNT could therefore alleviate the pain of neuropathies and various types of headache with neurogenic inflammation components. Another suggested site of analgesic action was the postganglionic sympathetic nerve ending that uses norepinephrine and ATP (adenosine tri-phosphate) as transmitters, where BoNT could be analgesic in cases of sympathetically maintained pain [37].

BoNT and Chronic Pain

Chronic pain is an altered state of central nervous system responsiveness due to repeated painful stimuli caused by inflammation, tissue or peripheral nerve damage that can severely impact patient's quality of life. The mechanism for peripheral sensitisation includes increased sensitivity of existing receptors or increased surface expression of additional receptors (capsaicin TRPV1 and purinergic P2X3) implicating a SNARE mediated vesicle-fusion pathway in protein kinase C potentiation [38-39].

The review by Casale and Tugnoli in 2008 already supports the 'off label' use of BoNT in the management of chronic pain. Although the pain-reducing effect of BoNT is mainly due to its ability to block acetylcholine release at the synapse, in vitro and in vivo data show that BoNT can also provide benefits related to effects on cholinergic control of the vascular system, autonomic function, and cholinergic control of nociceptive and antinociceptive systems. Furthermore, the authors also suggest that BoNT may influence the peripheral and central nervous systems [40].

Bont-Induced Alterations of Nociceptor Function in Sensory Neurons

Recent experimental studies are elucidating further the underlying mechanisms of analgesic effects of BoNT. The biochemical mechanism involving enzymatic blockade of neurotransmitter release however, is comparable between the motor nerve and the sensory nerve. The toxin was reported to block substance P release in vitro, to block the release of glutamate and prevent c-fos expression *in vivo* (gene involved in neuronal stimuli). These experimental data suggest that BoNT suppresses the secretion of neurotransmitters by directly decreasing peripheral sensitization and indirectly decreasing central sensitization which explains its effects on migraine pain by action on these two pathways [25].

BoNT A inhibits inflammatory pain in the rat formalin model, by direct inhibition of formalin-induced glutamate and also dose dependently reduced the number of formalin-induced Fos-like immunoreactive cells in the dorsal horn of the spinal cord. In addition, BoNT significantly inhibited the excitation of wide dynamic range neurons of the dorsal horn in Phase II of the formalin response, confirming that the toxin inhibits peripheral sensitization in this model, which also leads to an indirect reduction in central sensitization [41].

In addition, inhibition by BoNT of the evoked release of pain peptides from sensory neurons in the rat further confirmed their anti-nociceptive potential [42].

The review by Pavone and Luvisetto in 2010 concluded that BoNT/A induces long-term analgesic effects in many different pain models (from inflammatory to neuropathic) which is in good agreement with Aoki and Meng publications. Experimental data altogether support the hypothesis that the antinociceptive effect of BoNT/A can be attributed to the direct inhibition of peripheral sensitisation and to indirect inhibition of central sensitisation by prevention of the release of neuromodulators at peripheral nociceptive terminals.

The authors also suggest different mechanism of action of BoNTs, depending on the serotype and route of administration, which is extremely important to take into account in a therapeutic perspective [43].

Effects of BoNT-B on neurotransmitter release from spinal primary afferent sensory fibres and subsequent effects on nociception were studied in vivo in mice. A single intrathecal injection of BoNT-B 0.5 U blocked transmitter release from spinal primary afferents, attenuated inflammatory nociceptive response and spinal nerve injury-induced neuropathic pain in the absence of motor impairment.

These effects also correlated with BoNT-B cleavage of VAMPI/II protein in tissue lysate. BoNT-B also induced a reduction in phase 2 of formalin-evoked flinching behavior over 30 days after injection. These effects were reversed by co-injection of BoNT-B and anti-BoNTB antibody confirming the ability of BoNT-B to regulate spinal nociceptive processing [44].

Novel observations demonstrated bilateral reduction of pain after unilateral toxin injection, and proposed retrograde axonal transport, presumably in sensory neurons. Matak et al. in 2011 showed reduced formalin-induced pain after both peripheral and intraganglionic BoNT type A injection. Cleaved SNAP-25 was found in medullary dorsal horn (spinal trigeminal nucleus) 3 days following the peripheral treatment, even with low dose applied (3.5 U/kg). Prevention of BoNT antinociceptive effect by colchicine confirmed that axonal transport of Botulinum type A occurs via sensory neurons and is directed to sensory nociceptive nuclei in the CNS [45].

Although, there is increasing evidence that BoNT A may affect sensory nociceptor fibers as shown by experimental studies above, the expression of its receptors in clinical pain states, and its effects in human sensory neurons, are largely unknown. Immunostaining of synaptic vesicle protein subtype SV2A (a receptor for BoNT A) in a range of clinical tissues indicated increased SV2A levels in injured nerve fibers, including nerve fibers from idiopathic detrusor over-active muscle and painful bladder syndrome. BoNT A and E treatment of cultured human dorsal root ganglion neurons induced a reduction of capsaicin receptor (TRPV1) associated with dose-related inhibition of capsaicin responses. This study indicated that BoNT A and E may affect receptor expression and nociceptor function in sensory neurons in humans [46].

Clinical trials on the safety and efficacy of BoNT/A from 122 study sites in North America and Europe established its efficacy in controlling pain associated with chronic migraine but full mechanism of action remained to be established in humans [47]. BoNT (onabotulinumtoxinA, Botox®) was recently approved for the treatment of chronic migraine pain which has increased the interest in understanding the mechanism by which a peripherally administered therapeutic protein can reduce chronic pain. Non-clinical and clinical results show that BoNT reduces the symptoms associated with chronic pain by reducing local inhibition of neuropeptide release and by indirect reduction of central sensitisation [34].

Neuroprotective Effects of Botulinum Toxins in Epilepsy, Dystonia and Focal Cerebral Ischemia

In vitro studies have demonstrated the effects of BoNTs on the release of several neurotransmitters from central neurons. More recent *in vivo* studies have provided the first characterization of the effects of BoNT/E on CNS neurons in vivo. Intra-hippocampal injection of BoNT/E inhibits glutamate release and spike activity of pyramidal neurons in the rat resulting in significant inhibition of seizure activity in experimental models of epilepsy [48].

More than one-third of epilepsy patients are estimated to have pharmacoresistant epilepsy and surgical removal of the temporal lobes is not always applicable because of the presence of secondary foci and localization of primary foci in the deep brain. Delivery of BoNTs directly into the seizure focus in the brain is currently being investigated in animal models. Several reports have demonstrated that hippocampal injection of BoNT in the rat model of epilepsy prevents neuronal loss and reduces chronic seizures. Botulinum toxin was also shown to inhibit glutamate release from the hippocampal synaptosomes. These recent reports provide a plausible mechanism of action of BoNT in the CNS and suggest the possibility of its therapeutic application to human epilepsy [27].

More than an additional therapy, the use of botulinum toxins is opening new insights into the pathophysiology of movement disorders. Central nervous system changes were observed after botulinum toxin treatment of cervical and hand dystonia patients. All patients abnormal hemispheric asymmetry resolved 4 weeks after the same patients were treated with intramuscular injections of toxin. The use of botulinum toxin revealed unexpected systems-level abnormality in dystonia, which may lead to new treatment strategies [49]. Likewise, cases of patients with combined trigeminal neuralgia and hemifacial spasm improved with botulinum toxin treatment. The beneficial effect of botulinum toxin in both disorders supported the idea of this toxin having a central mechanism of action that acts by controlling neuronal hyperactivity in the brain stem, as well as its peripheral action [50].

In addition, accumulation of mediators such as glutamate is suspected to contribute to neuronal damage after an ischaemic insult and neuroprotective potential of BoNT has been proposed. Administration of BoNT/E in a rat model of focal ischaemia consistently reduced the loss of pyramidal neurons at 24 h. The neuroprotective effect of BoNT/E, however, was no longer significant at 6 weeks. These data provided evidence that blockade of synaptic transmitter release with BoNT delays neuronal cell death following focal brain ischaemia, but also underline the importance of assessing long-term neuroprotection [51].

New evidence continues to emerge today for central effects of BoNTs. In vivo work in an experimental model of trigeminal neuropathy suggests long-term beneficial effect of BoNT/A in craniofacial pain associated with neurogenic inflammation. Effects of BoNT/A on pain and dural extravasation in this model were dependent on axonal transport through sensory neurons, as evidenced by colchicine injections completely abolishing BoNT/A effects. Bilateral effects of BoNT/A associated with dependence on retrograde axonal transport confirm the presence of a central action site [52].

Designing Botulinum Toxins Look Alike As Targeted Neuronal Silencing Agents

To date, data gained on native BoNTs have been applied worldwide for treating numerous human conditions due to overactivity of nerves supplying various organs. The advantages of native BoNT include very long duration, high potency and localized action but possible side-effects of BoNT/A may lead to paralysis away from the injection site.

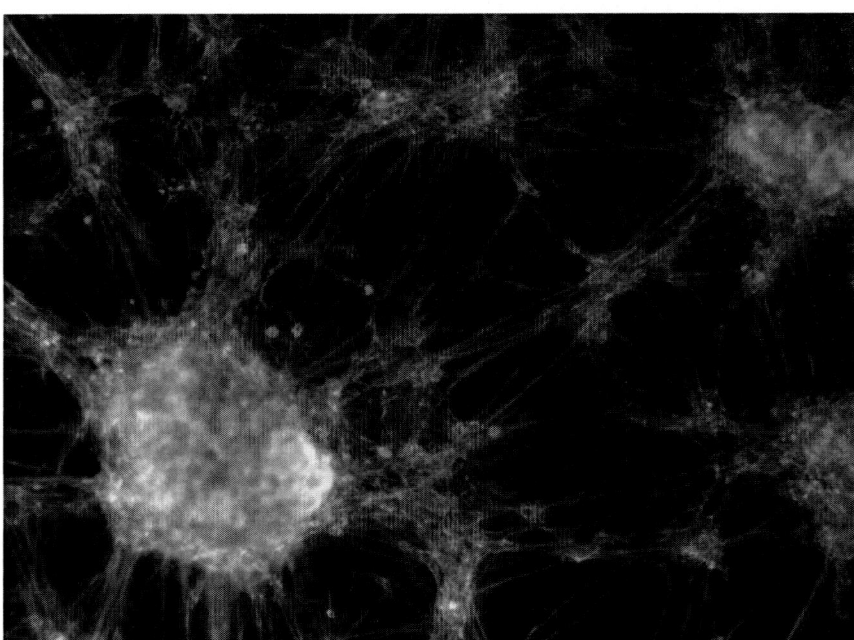

Figure 3. Uptake of re-assembled botulinum toxin BiTox (red) into synaptic vesicles of SH-SY5Y cells differentiated into neurons. (BiTox consists of the same three domains of the native BoNT/A with the SNARE stapling system between translocation and binding domains, Ferrari 2011).

A recent advance is the successful engineering of a chimera from two BoNTs to acquire the capability of targeting sensory neurons, with resultant inhibition of the release of a pain mediator. This novel recombinant protein blocks the exocytotic response triggered by a stimulant of nociceptive C fibres [9]. The capability of re-targeting a more active moiety to sensory neurons, with resultant inhibition of the release of a pain mediator suggest their potential application in the many cases of chronic pain that do not respond to existing drugs [53]. The proof of principle study by Wang et al. showed that new protein engineered toxins can be tailored for specific applications by targeting specific populations of neurons or secretory cells. The EA chimera was binding successfully to sensory neurons and blocking pain peptides release while BoNT type E alone was unable to bind to neurones [54].

In addition to the re-engineering of BoNT, a new "protein-stapling" technology was recently developed which allows re-assembly of BoNT/A from two separate fragments.

This new technology allowed safe production of a less paralytic version of a neuronal silencing agent. The re-assembled toxin exhibited a potent inhibition of CNS function but with no systemic toxicity after intra-peritoneal injection. This unique technology may also represent a new tool for the development of safer neuronal modulating agents needed in neuroscience research and medical applications [55].

Conclusion

Since 1980, when it was first approved for the treatment of strabismus, therapeutic indications for BoNT have been increasing exponentially to reach more than hundred indications to date. Licensed formulations of BoNT have been used in a wide range of conditions such as cervical dystonia, cerebral palsy, migraines and hyperhidrosis in addition to its worldwide use in beauty clinics.

Unforeseen beneficial effects have slowly emerged alongside its increased clinical use including pain relief in dystonias, chronic migraine or neurogenic inflammation. New studies demonstrate additional protective effects in experimental models of epilepsy or focal brain ischemia. Nevertheless, clinical studies elucidating central changes after BoNT treatment are still required for a better understanding of non-peripheral effects of BoNT on neuronal communication and modulation of vesicle trafficking activity.

There is an obvious need for new anti-nociceptive drugs to treat pain syndromes such as neuropathic pain and headache since a significant proportion of patients do not respond to medications currently available. In addition to native toxin, botulinum toxins “look alike” produced by re-targeting of recombinant chimera to nociceptive neurons or by "protein-stapling" technology of re-assembled BoNT from two separate fragments are opening new opportunities for chronic pain management using safer and more precisely targeted neuronal silencing agent.

Acknowledgment

I wish to thank Dr. Thea Sesardic for helpful suggestions.

References

Erbguth, R. J. and Naumann, M. (1999). Historical Aspects of Botulinum Toxin: Justinus Kerner (1786-1862) and the “Sausage Poison”. *Neurology, 53(8)*, 1850-1853.

Van Ermengem, E. (1979). Classics in infectious diseases. A new anaerobic bacillus and its relation to botulism. Originally published as "Ueber einen neuen anaëroben Bacillus und seine Beziehungen zum Botulismus" in Zeitschrift für Hygiene und Infektionskrankheiten 26: 1-56, 1897. A New Anaerobic Bacillus and Its Relation to Botulism. *Reviews of Infectious Diseases, 1(4)*, 701-719.

Bengtson IA. (1924). *Studies on organisms concerned as causative factors in botulism.* In: Office GP, editor. Hyg Lab Bull. Washington D. C.: U. S. Public Health Service.

Horowitz, BZ. (2012). History of Botulism. In: *Clostridium Botulinum: A Spore Forming Organism and a Challenge to Food Safety*, C.Rasetti-Escargueil and S. Surman Lee, New-York: Nova Publishers.

Lacy, D. B., Tepp, W., Cohen, A. C., DasGupta, B. R. and Stevens, R. C. (1998). "Crystal structure of botulinum neurotoxin type A and implications for toxicity", *Nat. Struct. Biol., 5(10),* 898-902.

Schiavo, G., Matteoli, M., and Montecucco, C. (2000). "Neurotoxins affecting neuroexocytosis", *Physiol Rev., 80,* 2, 717-766.

Simpson, L. L., (1981). The origin, structure, and pharmacological activity of botulinum toxin. *Pharmacol. Rev. 33*, 155-188.

Arnon, S. S., Schechter, R., Inglesby, T. V., Henderson, D. A., Bartlett, J. G., Ascher, M. S., Eitzen, E., Fine, A. D., Hauer, J., Layton, M., Lillibridge, S., Osterholm, M. T., O'Toole, T., Parker, G., Perl, T. M., Russell, P. K., Swerdlow, D. L., Tonat, K., (2001). Botulinum toxin as a biological weapon: medical and public health management. *JAMA 285*, 1059-1070.

Dolly, J. O., Lawrence, G. W., Meng, J., Wang, J., Ovsepian, S. V., (2009). Neuro-exocytosis: botulinum toxins as inhibitory probes and versatile therapeutics. *Curr. Opin. Pharmacol., 9*, 326-335.

Hanchanale VS, Rao AR, Martin FL, Matanhelia SS. (2010). The unusual history and the urological applications of botulinum neurotoxin. *Urol Int. 85(2)*, 125-30.

Naumann M, Albanese A, Heinen F, Molenaers G, Relja M. (2006). Safety and efficacy of botulinum toxin type A following long-term use. *Eur J Neurol, 13 (Suppl 4)*, 35-40.

Simpson, L. L., (1973). The interaction between divalent cations and botulinum toxin type A in the paralysis of the rat phrenic nerve-hemidiaphragm preparation. *Neuropharmacology 12*, 165-176.

Simpson, L. L., (1974). Studies on the binding of botulinum toxin type A to the rat phrenic nerve-hemidiaphragm preparation. *Neuropharmacology 13*, 683-691.

Binz T, Rummel A. (2009). Cell entry strategy of clostridial neurotoxins. *J. Neurochem. 2009, 109(6)*, 1584-95.

Montal M., (2010). Botulinum neurotoxin: a marvel of protein design. *Annu Rev Biochem., 79,* 591-617.

Smith, S. M., Renden, R., von, G. H., (2008). Synaptic vesicle endocytosis: fast and slow modes of membrane retrieval. *Trends Neurosci. 31*, 559-568.

Humeau, Y., Doussau, F., Grant, N. J., Poulain, B., (2000). How botulinum and tetanus neurotoxins block neurotransmitter release. *Biochimie, 82,* 427-446.

Ashton, A. C. and Dolly, J. O., (2000). A late phase of exocytosis from synaptosomes induced by elevated $[Ca2+]_i$ is not blocked by Clostridial neurotoxins. *J. Neurochem. 74(5)*, 1979-1988.

Davletov B, Bajohrs M, Binz T., (2005). Beyond BOTOX: advantages and limitations of individual botulinum neurotoxins. *Trends Neurosci. 8,* 446-52.

Pellizzari R, Rossetto O, Schiavo G, Montecucco C., (1999). Tetanus and botulinum neurotoxins: mechanism of action and therapeutic uses. *Philos Trans R Soc Lond B Biol Sci. 354(1381)*, 259-68.

Verderio C, Rossetto O, Grumelli C, Frassoni C, Montecucco C, Matteoli M. (2006). Entering neurons: botulinum toxins and synaptic vesicle recycling. EMBO Rep., 7(10), 995-9.

Verderio, C., Grumelli, C., Raiteri, L., Coco, S., Paluzzi, S., Caccin, P., Rossetto, O., Bonanno, G., Montecucco, C., Matteoli, M., (2007). Traffic of botulinum toxins A and E in excitatory and inhibitory neurons. Traffic. 8, 142-153.

Caleo, M., Antonucci, F., Restani, L., Mazzocchio, R., (2009). A reappraisal of the central effects of botulinum neurotoxin type A: by what mechanism? *J. Neurochem. 109(1)*, 15-24.

Caleo M, Schiavo G. (2009). Central effects of tetanus and botulinum neurotoxins. *Toxicon, 54(5)*, 593-9.

Aoki, K.R., (2003). Evidence for antinociceptive activity of botulinum toxin type A in pain management. *Headache, 43 Suppl. 1*, S9-15.

Ali, S. O., Thomassen, M., Schulz, G. M., Hosey, L. A., Varga, M., Ludlow, C. L., Braun, A. R., (2006). Alterations in CNS activity induced by botulinum toxin treatment in spasmodic dysphonia: an $H_2{}^{15}O$ PET study. *J. Speech Lang Hear. Res., 49,* 1127-1146.

Kato, K., Kohda, T., Kozaki, S., (2009). Application of botulinum neurotoxin in the treatment of epilepsy. *Brain Nerve, 61,* 939-948.

Currà A, Trompetto C, Abbruzzese G, Berardelli A. (2004). Central effects of botulinum toxin type A: evidence and supposition. *Mov Disord. 19 Suppl 8*, S60-4.

Abbruzzese G, Berardelli A. (2006). Neurophysiological effects of botulinum toxin type A. *Neurotox Res. Apr, 9 (2-3)*, 109-14.A.

Ceballos-Baumann AO, Sheean G, Passingham RE, Marsden CD, Brooks DJ. (1997). Botulinum toxin does not reverse the cortical dysfunction associated with writer's cramp: a PET study. *Brain, 120*, 571–82.

Palomar FJ, Mir P., (2012). Neurophysiological changes after intramuscular injection of botulinum toxin. *Clin Neurophysiol. 123(1)*, 54-60.

Antonucci F, Rossi C, Gianfranceschi L, Rossetto O, Caleo M., (2008). Long-distance retrograde effects of botulinum neurotoxin A. *J. Neurosci., 28(14),* 3689-96.

Restani L, Antonucci F, Gianfranceschi L, Rossi C, Rossetto O, Caleo M. (2011). Evidence for anterograde transport and transcytosis of botulinum neurotoxin A (BoNT/A), *J Neurosci., 31(44)*, 15650-9.

Aoki K.R., Francis J. (2011). Updates on the antinociceptive mechanism hypothesis of botulinum toxin A. *Parkinsonism Relat Disord. 17 Suppl 1,* S28-33.

Brin MF, Fahn S, Moskowitz C, Friedman A, Shale HM, Greene PE, Blitzer A, List T, Lange D, Lovelace RE, et al. (1987). Localized injections of botulinum toxin for the treatment of focal dystonia and hemifacial spasm. *Mov Disord. 2(4)*, 237-54.

Arezzo JC. (2002). Possible mechanisms for the effects of botulinum toxin on pain. *Clin J Pain.* 18(6 Suppl), S125-32.

Mense S. (2004). Neurobiological basis for the use of botulinum toxin in pain therapy. *J Neurol. 251 Suppl 1*, I1-7. Aoki K.R., (2005). Review of a proposed mechanism for the antinociceptive action of botulinum toxin type A. *Neurotoxicology, 6(5),* 785-93.

Woolf CJ, Salter MW. (2000). Neuronal plasticity: increasing the gain in pain. *Science, 288(5472)*, 1765-9.

Morenilla-Palao C, Planells-Cases R, Garcia-Sanz N, Ferrer-Montiel A. Regulated exocytosis contributes to protein kinase C potentiation of vanilloid receptor activity. (2004). *J Biol Chem, 279(24)*, 25665-72.

Casale R, Tugnoli V. Botulinum toxin for pain. *Drugs R D. 2008, 9(1)*:11-27.

Aoki K.R. (2005). Review of a proposed mechanism for the antinociceptive action of botulinum toxin type A. *Neurotoxicology, 26(5)*,785-93.

Meng J, Wang J, Lawrence G, Dolly JO. (2007). Synaptobrevin I mediates exocytosis of CGRP from sensory neurons and inhibition by botulinum toxins reflects their anti-nociceptive potential. *J Cell Sci. 120(Pt 16),* 2864-74.

Pavone F, Luvisetto S. (2010). Botulinum neurotoxin for pain management: insights from animal models. *Toxins, 2*, 2890-913.

Huang PP, Khan I, Suhail MS, Malkmus S, Yaksh TL. (2011). Spinal botulinum neurotoxin B: effects on afferent transmitter release and nociceptive processing. *PLoS One., 6(4)*:e19126.

Matak I, Bach-Rojecky L, Filipović B, Lacković Z. (2011). Behavioral and immunohistochemical evidence for central antinociceptive activity of botulinum toxin A. *Neuroscience. 186,* 201-7.

Yiangou Y, Anand U, Otto WR, Sinisi M, Fox M, Birch R, Foster KA, Mukerji G, Akbar A, Agarwal SK, Anand P. (2011). Increased levels of SV2A botulinum neurotoxin receptor in clinical sensory disorders and functional effects of botulinum toxins A and E in cultured human sensory neurons. *J Pain Res., 4*, 347-55.

Schoenen J. (2010). Botulinum toxin in headache treatment: finally a promising path? *Cephalalgia, 30(7)*, 771e3.

Bozzi Y, Costantin L, Antonucci F, Caleo M. (2006). Action of botulinum neurotoxins in the central nervous system: antiepileptic effects. *Neurotox Res. 9(2-3)*, 197-203.

Blood AJ, Tuch DS, Makris N, Makhlouf ML, Sudarsky LR, Sharma N. *(2006).* White matter abnormalities in dystonia normalize after botulinum toxin treatment. *Neuroreport, 17(12),* 1251-5.

Boscá-Blasco ME, Burguera-Hernández JA, Roig-Morata S, Martínez-Torres I. (2006). Painful tic convulsif and Botulinum toxin, *Rev Neurol. 42(12)*, 729-32.

Antonucci F, Cerri C, Maya Vetencourt JF, Caleo M. (2010). Acute neuroprotection by the synaptic blocker botulinum neurotoxin E in a rat model of focal cerebral ischaemia, *Neuroscience, 69(1)*, 395-401.

Filipović B, Matak I, Bach-Rojecky L, Lacković Z. (2012). Central action of peripherally applied botulinum toxin type A on pain and dural protein extravasation in rat model of trigeminal neuropathy. *PLoS One. 7(1)*, e29803.

Foster KA, Adams EJ, Durose L, Cruttwell CJ, Marks E, Shone CC, Chaddock JA, Cox CL, Heaton C, Sutton JM, Wayne J, Alexander FC, Rogers DF. (2006). Re-engineering the target specificity of Clostridial neurotoxins - a route to novel therapeutics. *Neurotox Res., 9(2-3)*, 101-7.

Wang J, Meng J, Lawrence GW, Zurawski TH, Sasse A, Bodeker MO, Gilmore MA, Fernandez-Salas E, Francis J, Steward LE et al. (2008). Novel chimeras of botulinum neurotoxins A and E unveil contributions from the binding, translocation, and protease domains to their functional characteristics. *J. Biol. Chem*, 283, 16993-17002.

Ferrari E, Maywood ES, Restani L, Caleo M, Pirazzini M, Rossetto O, Hastings MH, Niranjan D, Schiavo G, Davletov B. (2011). Re-Assembled Botulinum Neurotoxin Inhibits CNS Functions without Systemic Toxicity. *Toxins (Basel), 3(4)*, 345-55. R.

In: Botulinum Toxin
Editors: Fausto E. Marino and Sebastiano De Luca
ISBN: 978-1-62257-185-7

Chapter III

Aesthetic Indications of Botulinum Toxin: A Review of the Efficacy and Safetyof Currently Available Products

***Luca A. Dessy*[1], *Nefer Fallico*[1], *Marco Mazzocchi*[2] *and Nicolò Scuderi*[1]**

[1] Department of Plastic and Reconstructive Surgery, Sapienza University of Rome, Rome, Italy

[2] Department of Plastic and Reconstructive Surgery, University of Perugia, Perugia, Italy

Abstract

Botulinum toxin (BTX) is a powerful neurotoxin produced by the bacterium Clostridium botulinum. Its mechanism of action consists of the induction of a temporary muscle paralysis through the inhibition of the release of acetylcholine into the synaptic cleft. The finding that BTX smoothed facial lines when used therapeutically prompt researchers to study the toxin effect on glabellar lines. In fact, the onset of glabellar lines is closely related to the underlying muscles' activity. Nowadays, BTX injections for glabellar lines are commonly performed and an increasing number of formulations is becoming available.

This chapter provides a comparative evaluation of currently available botulinum toxins for the treatment of glabellar lines. Toxins have been divided into three groups: those that have obtained approval for cosmetic use (approved treatments), those that have not (off-label treatments), and those that are still under investigation (experimental treatments). The basic similarities and differences between available botulinum toxins are examined, and the pharmacokinetics and dosing are summarized.

Approved treatments are composed of a sole serotype of botulinum toxin. Indeed BTX type A (BTX-A) is the only approved for the treatment of glabellar lines. The main difference between BTX-A formulations is their molecular weight: original formulations are made of a BTX-A complex of 900 kDa while the latest is a 150 kDa BTX-A free from complexing proteins. With reference to the dosage, the main aspect emerging from this comparison is that despite considering the same serotype each formulation possesses distinctive characteristics due to its purification and manufacturing processes. As such, BTX formulations are not interchangeable. The methods of evaluation used for each toxin are investigated and the safety and efficacy data reported in the literature are reviewed. Major adverse events are uncommon while minor adverse events, such as headache, blepharoptosis, and injection-site reactions, are relatively frequent but transient.

Off-label treatments consist of botulinum toxins that are approved for therapeutic applications but are used off-label for cosmetic indications. Currently, the only formulation available on the market is BTX type B. Published evidence on its safety and efficacy is extremely scarce.

Experimental treatments are composed of BTX products under investigation or recently launched for aesthetic applications. These formulations are designed to overcome some of the limitations present in the older generation of products. However, some of these toxins may be easily purchased via the Internet but do not have any license neither for cosmetic nor therapeutic indications. For this reason, such products are potentially a severe health risk for patients.

Origin and Mechanism of Action

Botulinum toxin (BTX) is a powerful neurotoxin produced by different strains of the gram-positive bacterium named Clostridium botulinum and is labeled in seven serotypes (A-G). Botulinum toxin is a zinc-dependent endopeptidase composed of a light (50 kDa), and a heavy (100 kDa) chain linked by disulfide bonds.

Botulinum toxins work by inhibiting the release of acetylcholine into the synaptic cleft, which results in temporary muscle paralysis. Upon injection of botulinum toxin in a target tissue, the heavy chain binds to specific glycoprotein structures on the plasma membrane of the nerve cell, which invaginates around the toxin-receptor complex. The toxin-receptor complex is then internalized into the synapse. The light chain of the botulinum toxin is then released into the cytoplasm, where it binds to a target protein, which varies among botulinum toxins' serotypes. Botulinum toxins A and E cleave synaptosome-associated proteins of 25 kDa (SNAP25), while botulinum toxins B, D, F and G cleave vesicle-associated membrane protein (VAMP), also known as synaptobrevin. The light chain produces a proteolytic cleavage of the target protein preventing the fusion of the acetylcholine vesicle on the cellular membrane. If the target tissue is a muscle, the cascade of events produces a chemical denervation resulting in muscle paralysis. Results of botulinum toxin injections lasts 2 to 6 months and then diminish, when the protein complex turnover is restored.

Aesthetic Indications

Ophthalmologists in the early 80's were the first to use botulinum toxin therapeutically and soon after it was observed that its application smoothed facial lines, which prompted researchers to investigate the aesthetic indications for this treatment. By 1992, studies established that botulinum toxin could be used to diminish the appearance of glabellar lines. Clinical evidence suggest that it can be applied to other hyperkinetic facial lines (crow's feet, horizontal forehead lines, nasolabial folds) with good results but at the moment the only approved use is for glabellar lines.

Evaluation

The assessment of results after botulinum toxin injections for cosmetic purposes can be performed by both subjective and objective methods.

Photography is the most common evaluation technique. It allows a comparison between the pre- and post-treatment aspect of the patients, not only with regards to the glabellar area, but also to any previous asymmetry that the patient may notice only afterwards. Visuals consist of a band of both eyes caught when relaxed and frowning.

Patient satisfaction is the most important outcome in facial aesthetic surgery and it has therefore been included in several major clinical trials [1]. Patient satisfaction is often rated by number on an ordinal scale, such as the Facial Lines Treatment Satisfaction Questionnaire (FLTSQ) [2,3].

Moreover, patients attitude on beauty and body can be assessed using scales of the Freiburg Questionnaire on Aesthetic Dermatology and Cosmetic Surgery (FQAD), and the Freiburg Life Quality Assessment core (FLQA-C) [4].

Investigators have recently begun to assess other outcomes deemed important to patients undergoing aesthetic treatments. The outcomes include the Self-Perception of Age (SPA) measure and the Facial Line Outcomes (FLO) questionnaire. On the FLO questionnaire, patients rate the extent to which facial lines make them look tired, stressed, or angry when that does not express their feelings [5].

A more objective measure of treatment outcomes is the glabellar line severity scale (GLSS). It is a 4-point scale that grades wrinkles and lines from 0 (none) to 3 (severe), as evaluated by a trained observer, or independent assessor.

Performing a negative cast of superficial skin texture with the silicone replica technique [6,7] or a direct in vivo no-touch measurement of skin surface [8,9] before and after treatment can give an objective assessment. In the negative cast, araldite contrareplicas allow professionals to obtain a permanent positive unalterable representation of the skin surface. In the no-touch measurement, the system provides skin topography directly generating 3D images reproducing the skin surface. Subsequently, skin surface parameters are studied by optical profilometry or scan electron microscopy (only on araldite contrareplicas) by using a specific computer software.

Available Formulations

Over the last two decades, BTX became a popular treatment because of its noticeable results and minimally invasive nature. As demand for this treatment continues to increase, a growing number of formulations are becoming available.

For better clarification, we have divided available toxins into three groups: those that have obtained approval for cosmetic use (approved treatments), those that have not (off-label treatments), and those that are still under investigation (experimental treatments).

Approved Treatments

The only approved treatments are composed of serotype A of botulinum toxin (BTX-A). At present, there are three commercial formulations of BTX-A approved for cosmetic indications: Onabotulinum toxin A (OnT-A), Abobotulinum toxin A (AbT-A), Incobotulinum toxin A (InT-A) or NT-201 (Table.1).

Onabotulinum Toxin A (Botox® Cosmetic/Vistabel®/Vistabex®)

OnT-A, under the commercial name Botox® Cosmetic (Allergan Inc., Irvine, CA, USA.), was approved by the U.S. Food and Drug Administration (FDA) in 2002 for the temporary improvement in appearance of moderate to severe glabellar lines. In Italy, it was authorized with the name of Vistabex® by the Ministry of Health in 2004. In the U.K. and several other European countries, this product was later licensed under the brand name Vistabel®.

OnT-A is produced from a culture of Hall strain of *Clostridium botulinum* and purified by a series of acid precipitations to a crystalline complex containing the toxin and other proteins. The complex is lyophilized and presents as a sterile powder. Shantz was the first to purify OnT-A in crystalline form but it was Scott who adapted it for clinical use.

Originally, it was only available as a 100 unit vial but, in order to gain European regulatory approval, a vial containing 50 units was produced for cosmetic use.

Vials must be kept at a temperature between 2°C and 8°C. Before injection, the powder must be diluted with 0.9% normal saline. The package insert recommends using a sterile non-preserved saline, although it has been demonstrated that preserved saline does not decrease efficacy. The recommended dilution ranges from 1-4 mL per 100 units vial (0.5-2 mL per 50 units vial).

Dilution can be adjusted according to the desired degree of migration. For treatment of glabellar lines, the authors prefer to use 1.5 mL of sterile normal saline for the 50 U vial to prepare a solution of 3.3 U/0.1 mL (33 U in a 1 mL insulin syringe).

Table 1.

Commercial name	BOTOX® Cosmetic/ Vistabel®/ Vistabex®	Dysport®/ Azzalure®	Xeomin®/ Bocouture®	Myobloc®/ NeuroBloc®	PurTox®	Neuronox®	-----/ Prosigne®	-----
International Nonproprietary Name (INN)	Onabotulinum toxinA	Abobotulinum toxinA	Incobotulinum toxinA	Rimabotulinum toxinB	-----	-----	CBTX-A	CNBTX-A
Company	Allergan Inc.	Medicis Inc./ Galderma	Merz Pharmaceuticals	Solstice Neurosciences Inc./ Eisai Co., Ltd.	Mentor Corporation	Medy-Tox Inc.	Lanzhou Biological Products Institute	Nanfeng Medical Science and Technology Development Co.,Ltd.,
Cosmetic Approvals	North America, Europe	North America, Europe	Argentina, Mexico, Germany, U.K., France and Austria	Some European countries, USA, Canada	None	None	China, Brasil	None
Indications	Glabellar lines	Glabellar lines	Glabellar lines (Germany and Argentina); hyperkinetic facial lines (Mexico)	Cervical dystonia; off-label use for glabellar lines	Phase 3 for glabellar lines	Pre-clinical studies only	-----	-----
Type	Type A-Hall strain	Type A	Type A-Hall strain	Type B	Type A-Hall strain	-----	-----	-----
Active Substance (molecular weight)	Botulinum toxin type A complex (900kDa)	Botulinum toxin type A complex (500-900kDa)	Botulinum toxin type A, free from complexing proteins (150kDa)	Botulinum toxin type B complex (700kDa)	Botulinum toxin type A, free from complexing proteins (150kDa)	-----	-----	-----
Mode of Action	SNAP-25	SNAP-25	SNAP-25	VAMP	SNAP-25	-----	-----	-----
Pharmaceutical Form	Powder dissolved in solution for injection	Powder dissolved in solution for injection	Powder dissolved in solution for injection	Solution	-----	-----	-----	-----

Table 1. (Continued)

Commercial name	BOTOX® Cosmetic/ Vistabel®/ Vistabex®	Dysport®/ Azzalure®	Xeomin®/ Bocouture®	Myobloc®/ NeuroBloc®	PurTox®	Neuronox®	-----/ Prosigne®	-----
Storage	2-8°C or < -5°C	2-8°C	Up to 25°C	2-8°C do not freeze	-----	-----	-----	-----
Units/vial	100U.I. 50U.I.	500s.U. 300s.U. 125s.U.	50U.I.	2.500; 5.000; 10.000	-----	100	-----	-----
Re-constitution	0.9% sodium chloride	0.9% sodium chloride	0.9% sodium chloride	Prepared solution, dilutable	0.9% sodium chloride	-----	-----	-----
Excipients	Human albumin, Sodium chloride	Human albumin, Lactose monohydrate	Human albumin, Sucrose	-----	-----	-----	Bovine gelatin protein	-----
Strength of Action (BTX-A:Product)	1:1	1:2 - 1:4 (approximate)	1:1	1:50 - 1:100	1:1.5 (?)	-----	-----	-----

The reconstituted vial must be kept at a temperature between 2°C and 8°C. Although the package insert indicates that reconstituted OnT-A should be used within 4 hours, one study found no loss of activity at 6 hours, a 44% loss after 12 hours and a 70% loss at 1-2 weeks. As such, most practitioners discard unused reconstituted vials after 1-7 days.

Injection Technique

In order to perform the injections, a 30-gauge 1-inch needle is used.

The glabellar area is injected in 5 sites with 3-6 units at each site for an average total dose of 15-25 units. The 5 sites of injection are 1 in the procerus and 2 in each side of corrugator supercilii (one superior to the medial canthus 1 mm above the eyebrow and the other one more laterally approximately 4 mm above the eyebrows) (Figure 1). The needle is advanced to the periosteum and then withdrawn slightly to place the needle in the muscle. According to the authors' opinion, it is preferable to start treatment with the lowest dose in order to test patient responsiveness and, if required, to increase botulinum units in succeeding treatment sessions.

Figure 1. Schematic drawing showing botulinum toxin A injection sites in the glabellar area.

Some patients may need evaluation for medial recruitment of the mid brow caused by hyperfunctional orbicularis oculi fibers. In these patients, an additional 3-6 units can be injected 1 cm above the supraorbital prominence vertical to the midpoint of the eyebrow to avoid resultant accentuation of eyebrow arching.

Injection techniques often vary among practitioners and over time. Patients usually notice a clinical effect 1-3 days following injection, and the effect is maximal by 1-2 weeks. Some diffusion of toxin occurs.

Evaluation Techniques

The effect of OnT-A treatment on glabellar rhytides has been assessed with several methods, including photography, patient satisfaction, patient-reported outcomes and silicone skin replica technique [1,3,11-15].

The FLO questionnaire has been used to assess the effect of OnT-A treatment on glabellar lines. The glabellar-specific FLO questionnaire investigates 7 items, including the SPA item [16]. FQAD and FLQA-C scales have been used to evaluate patients' attitude on beauty and body along treatment outcome satisfaction, with the intent to predict adherence to the OnT-A treatment [17].

OnT-A treatment on glabellar lines has been also investigated with the silicone skin replica technique, which shows objective skin texture changes.

Side Effects

Adverse reactions may be related to treatment, injection technique or both.

In general, adverse reactions occur within the first few days following injection and are transient.

The frequency of adverse reactions is defined as follows: very common ($\geq$ 1/10); common ($\geq$ 1/100, <1/10); uncommon ($\geq$ 1/1,000, <1/100); rare ($\geq$ 1/10,000, <1/1,000); very rare (<1/10,000).

In the patient information leaflet, the following reactions are described as common adverse effects: headache, eyelid ptosis, erythema, localised muscle weakness, face pain. Reported uncommon adverse effects are: infection, anxiety, paresthesia, dizziness, blepharitis, eye pain, visual disturbance, nausea, oral dryness, skin tightness, oedema (face, eyelid, periorbital), photosensitivity reaction, pruritus, dry skin, muscle twitching, flu syndrome, asthenia, fever.

Patients may develop neutralizing antibodies. In these occurrences, serum can be tested, although this is rarely performed outside research settings. Alternatively, a patient's physiologic response can be evaluated with a single injection of 15 units into the frontalis muscle on one side. It is still unknown whether neutralizing antibodies resolve over time and, consequently, whether attempts at reinjection should be made after a prolonged period. To reduce these occurrences, the use of the lowest dose of toxin possible and the avoidance of reinjection within 1 month appear prudent.

Abobotulinum Toxin A (Dysport®/Azzalure®)

AbT-A was approved for cosmetic use in April 2009 by the FDA under the commercial name of Dysport® (Medicis Aesthetics Inc., Scottsdale, AZ, USA) [18]. In Europe, AbT-A is marketed under the name of Azzalure® (Galderma, Düsseldorf, Germany).

This formulation of type A neurotoxin is available in Europe, North America and a few other countries.

AbT-A is produced by purification using a column separation method, whereas OnT-A is purified by repeated precipitation and redissolution. The distinct processes used for purification accounts for some differences in the surrounding multi-protein complex that is formed around the neurotoxin; however, the clinical significance of this variation is still unclear [19].

Units of AbT-A are expressed in Speywood units (s.U). Dysport® is distributed in 300 or 500 s.U. vials, Azzalure® in 125 s.U. vials.

Vials must be stored at a temperature between 2°C and 8°C.

Before injection, the powder must be diluted with sterile 0.9% unpreserved sodium chloride.

The 500s.U. vials are reconstituted with 5mL of solution (10 s.U./0.1 mL). The 300 s.U. vials are reconstituted with 2.5 mL of solution (10 s.U./0.08 mL). It may also be reconstituted with 1.5 mL solution (10s.U./0.05 mL). The 125 s.U. vials are reconstituted with 0.63 mL of normal saline (10 s.U./0.05 mL).

Once reconstituted, the vial must be kept at a temperature between 2°C and 8°C. The package insert indicates that reconstituted AbT-A should be used within 4 hours.

Injection Technique

The injection scheme of AbT-A for the treatment of glabellar frown lines is the same of that of OnT-A (five point injection sites). The recommended dose is of 10 s.U. for each injection site for a total dose of 50 s.U. In a study on the treatment of glabellar frown lines, AbT-A demonstrated the optimum efficacy at a 50 s.U. dose [20].

Evaluation Technique

The methods of evaluation that have been used in order to assess the effect of AbT-A treatment on glabellar lines include photography, patient satisfaction and patient-reported outcomes [21-23].

The GLSS at maximum frown has been used to assess the effect of AbT-A treatment on glabellar lines and diaries kept by the patients were used to document onset of effect [24].

Bioequivalence

AbT-A doses are approximately three times the OnT-A doses, though bioequivalence conversion factors for the two treatments remain uncertain [25,26]. In fact, it has been shown that the conversion rate may be less than 1:3. Recent study revealed that 36 s.U. of AbT-A had a significantly longer duration of action in forehead wrinkles than 12 U of OnT-A. Nevertheless, AbT-A retains a tendency for greater adverse effects even with more equalized dosage ratios. These findings imply that OnT-A and AbT-A are not easily interchangeable.

Side Effects

Adverse reactions related to treatment or injection technique are usually of mild to moderate severity. They occur within the first week following injection and are transient.

According to the patient information leaflet, common adverse reactions are: headache, asthenopia, ptosis, eyelid oedema, lacrimation increase, dry eye, muscle twitching, injection site reactions.

Reported uncommon adverse reactions are: dizziness, visual disturbances, blurred vision, diplopia, pruritus, rash.

Incobotulinum Toxin A (Xeomin®/ Bocouture®)

The latest BTX-A formulation to receive approvals for cosmetic use is InT-A, under the commercial name of Xeomin® or Bocouture® [27].

InT-A is an uncomplexed botulinum toxin type A produced by Merz Pharma GmbH, Frankfurt, Germany.

InT-A is approved for the treatment of blepharospasm and cervical dystonia in Europe, Canada, Russia and Mexico under the trade name Xeomin® [28]. Cosmetic approval is available in Argentina for the temporary improvement of the appearance of glabellar lines, and in Mexico for the treatment of hyperkinetic facial lines. As Bocouture®, it recently received cosmetic approval in Germany, U.K., France and Austria for the treatment of glabellar lines. Release is planned for the rest of Europe and the U.S. in the near future [29].

InT-A is a highly purified formulation composed of BTX-A, which is free of complexing proteins [30]. Recent study demonstrated that the 150 kDa active neurotoxin is rapidly released from the 900 kDa complex under physiological conditions, suggesting that complexing proteins do not play an essential role in stabilizing the BTX-A complex after injection or minimizing the diffusion of the active component [31,32].

Because immunogenicity is largely dependent on foreign protein load, the fact that InT-A contains only the active neurotoxin with no additional clostridial protein signifies its immunogenic potential will be minimal.

Additional benefits offered by a highly purified formulation include improved handling characteristics, i.e. unreconstituted vials of InT-A powder are stable at room temperature for up to 36 months with no need for refrigeration, but must be kept under a temperature of 25°C [33].

In order to be injected, the reconstitution is performed with 0.9% sodium chloride. The package insert recommends dilution with sterile unpreserved saline. The recommended dilution is 1.25 mL per 50 units. The final dilution of InT-A is 40 units/mL.

Once reconstituted, the vial must be kept in a refrigerator at a temperature between 2°C and 8°C. The package insert indicates that reconstituted OnT-A demonstrates chemical and physical stability for 24 hours. However, from a microbiological point of view, the product should be used immediately.

Injection Technique

After reconstitution of InT-A (50 units/1.25 ml), the recommended injection volume of 0.1 ml (4 units) is injected into each of the 5 injection sites, according to the previously described scheme for OnT-A. The dose may be increased by the physician to up to 30 units if required by the individual needs of the patients, with at least 3-months interval between treatments. A 30-gauge 1-inch needle is used to perform the injections.

Bioequivalence

In terms of potency, InT-A appears to exhibit a 1:1 dose ratio when compared with OnT-A.

In studies that compared InT-A with OnT-A for therapeutic indications, no significant differences were detected in both safety and efficacy [34].

Study showed that efficacy is lower in patients older than 50 years compared to younger patients and in men compared to women.

Side Effects

As in OnT-A and AbT-A, undesirable effects are usually observed within the first week after treatment and are temporary and may be related to the active substance, the injection procedure or both.

Based on clinical experience, common undesirable effects are: headache, localized muscle weakness, sensation of heaviness. Uncommon undesirable effects are: influenza, infection, brow ptosis, vasovagal syncope, localized pain, tenderness, itching, swelling and/or haematoma, injection site reactions. It is interesting to note thatb compared to OnT-A and AbT-A, injection site reactions are less common.

Off-Label Treatments

Some botulinum toxins that are approved for therapeutic applications, such as botulinum toxin type B (BTX-B), are used for cosmetic purposes with off label indications [35,36].

The only approved BTX-B formulation has the non-proprietary name of Rimabotulinum toxin B (RiT-B) for the treatment of cervical dystonia [37,38]. However, several studies have proven the efficacy of RiT-B in the treatment of glabellar lines as well [39,40].

Rimabotulinum Toxin B (Myobloc®/NeuroBloc®)

This neurotoxin was originally developed by Elan Pharmaceuticals, but it is now produced by Solstice Neurosciences Inc.. Eisai Co., Ltd. holds the commercial rights to the product in Japan, Russia, European Union nations, as well as other countries. RiT-B is approved in Canada, US and Europe for cervical dystonia.

RiT-B is available as a solution with an acidic pH level of 5.5 to 5.6 and because of this elevated acidity, patients may find injections to be more painful than approved formulations. Moreover, RiT-B units are substantially less effective when compared with OnT-A. In fact, in cervical dystonia, the ratio is estimated to be approximately 50:1, whereas for glabellar frown lines, the ratio is 100:1. The clinical duration did not appear to be as long-lasting as OnT-A [41]. However, RiT-B does appear to exhibit a more rapid onset of action and a greater area of diffusion [42]. Associated adverse effects, such as dry mouth and dysphagia, are generally mild-to-moderate, transient, and more common at higher doses.

Experimental Treatments

A number of BTX-A products are currently in development or have recently been launched for aesthetic applications. These products have been specifically designed to overcome some of the limitations present in the older generation of products.

These include PurTox® (Mentor Corporation, Santa Barbara, CA, USA), and Neuronox® (Medy-Tox Inc., South Korea) as well as BTX-A preparations produced by the Lanzhou Biological Products Institute, Lanzhou, Gansu, China (marketed as CBTX-A) and the Nanfeng Medical Science and Technology Development Co. Ltd., Shijiazhuang, Hebei, China (marketed as CNBTX-A).

PurTox®

PurTox® is an uncomplexed BTX-A produced by Mentor Corporation, Santa Barbara, CA, USA, currently in its last stage of testing for its use in glabellar frown lines [43]. The units of PurTox® appear to be slightly less effective than OnT-A.

Neuronox®

Neuronox® is a BTX-A complex manufactured by Medy-Tox Inc. in South Korea.

Neuronox® is marketed in vials of 100 units. Published data on Neuronox® is extremely scarce and limited to pre-clinical studies [44]. However, information suggests that the clinical trial program for Neuronox® involves facial cosmetic application [45].

CBTX-A

CBTX-A is a BTX-A formulation produced by the Lanzhou Biological Products Institute, Lanzhou, Gansu, China. It is the only product that is approved in the People's Republic of China and it is marketed in Brazil as Prosigne®.

Unique to this treatment, a bovine gelatin protein, instead of human serum albumin, is added to the vial in order to prevent the toxin from sticking to its wall. Because of its bovine derivation, the gelatin utilized has the potential to induce allergic reactions, or possibly bovine spongiform encephalopathy.

CNBTX-A

CNBTX-A is a BTX-A formulation produced by the Nanfeng Medical Science and Technology Development Co., Ltd., Shijiazhuang, Hebei, China. CNBTX-A is neither licensed nor approved in any country although it can be purchased on the internet. An investigation found this product to contain significantly higher levels of botulinum neurotoxin than listed on the product's label, which could constitute a severe health risk for patients [46].

Discussion

Facial rhytides represent a widespread aesthetic concern. Facial lines are a result of a combination of intrinsic skin aging, ultraviolet damage, and repetitive action of underlying facial muscles;[47] the onset of glabellar lines is closely related to the movements of the underlying procerus, corrugators supercilii and orbicularis oculi muscles [48].

The presence of these lines when the face is at repose may give an erroneous impression of negative emotions. In addition, facial lines are considered problematic because they are perceived as a sign of aging [49,50].

Individuals have sought to improve their appearance throughout history and across virtually all cultures [51]. Nowadays, medical technology has made an increasing number of surgical and minimally invasive treatment options available to meet these desires and to maintain a more youthful appearance [52]. Botulinum toxins represent the foundation of minimally invasive aesthetic facial treatments. These begin with the use of botulinum toxin to smooth glabellar frown lines and expand to include other facial areas as well as combination use with other minimally invasive agents and procedures [53].

At present, there is a limited number of botulinum toxin formulations licensed for aesthetics procedures, though a number of products are currently under development. There is a substantial body of published evidence on the use of these approved treatments for facial enhancement, proving efficacy and safety. Minor adverse effects are relatively frequent but transient, whilst major adverse effects are rare.

The important aspect that has come about when considering approved botulinum toxins for cosmetic indications is that even with a given serotype, such as botulinum toxin type A, formulations are not interchangeable. This means that clinicians cannot apply the same dosage for all formulations, but must instead learn to use each product properly to provide optimal and safe outcomes for their patients.

Some botulinum toxins that are approved for therapeutic applications are used for cosmetic purposes with off label indications, thus without the approval of the health regulatory committees (i.e. USA FDA, Minister of Health) and without sufficient published evidence on safety and efficacy. Other types of toxins may be easily purchased via the internet, without having any license or approval for cosmetic or therapeutic indications in any country; these products must be considered unsafe and potentially a severe health risk for patients.

References

Fagien S, Carruthers JDA. A comprehensive review of patient-reported satisfaction with botulinum toxin type A for aesthetic procedures. *Plast. Reconstr. Surg.*, 2008;122(6):1915-1925.

Cox SE, Finn JC, Stetler L, Mackowiak J, Kowalski JW. Development of the Facial Lines Treatment Satisfaction Questionnaire and initial results for botulinum toxin type A-treated patients. *Dermatol. Surg.*, 2003;29(5):444-449.

Stotland MA, Kowalski JW, Ray BB. Patient reported benefit and satisfaction with botulinum toxin type A treatment of moderate to severe glabellar rhytides: results from a prospective open-label study. *Plast. Reconstr. Surg.*, 2007;120(5):1386-1393.

Sommer B, Zschocke I, Bergfeld D, Sattler G, Augustin M. Satisfaction of patients after treatment with botulinum toxin for dynamic facial lines. *Dermatol. Surg.*, 2003;29(5):456-460.

Carruthers J, Carruthers A. Botulinum toxin type A treatment of multiple upper facial sites: patient-reported outcomes. *Dermatol. Surg.*, 2007;33(1 Spec No):S10-S17.

Dessy LA, Mazzocchi M, Rubino C, Mazzarello V, Spissu N, Scuderi N. An objective assessment of botulinum toxin A effect on superficial skin texture. *Ann. Plast. Surg.*, 2007 May;58(5):469-473.

Caputo R, Monti M, Motta S, et al. The treatment of visible signs of senescence: the Italian experience. *Br. J. Dermatol.*, 1990;122 (suppl 35):97-103.

Lagarde JM, Rouvrais C, Black D, Diridollou S, Gall Y. Skin topography measurement by interference fringe projection: a technical validation. *Skin. Res. Technol.*, 2001;7:112-121.

Altmeyer P, Erbler H, Kromer T, Duwe HP, Hoffmann K. Interferometry: a new method for no-touch measurement of the surface and volume of ulcerous skin lesions. *Acta. Derm. Venerol.*, 1995;75:193-197.

Carruthers A, Carruthers J. Prospective, double-blind, randomized, parallel-group, dose-ranging study of botulinum toxin type A in men with glabellar rhytids. *Dermatol. Surg.*, 2005;31(10):1297-1303.

Grimes PE. A four-month randomized, double-masked evaluation of the efficacy of botulinum toxin type A for the treatment of glabellar lines in women with skin types V and VI. Poster presented at: the 62nd Annual Meeting of the American Academy of Dermatology, February 6–11, 2004; Washington, DC.

Ellis DAF, Chi PL, Tan AKW. Facial rejuvenation with botulinum. *Dermatol. Nurs.*, 1997;9(5):329-333.

Dessy LA, Mazzocchi M, Rubino C, Mazzarello V, Spissu N, Scuderi N. An objective assessment of botulinum toxin A effect on superficial skin texture. *Ann. Plast. Surg.*, 2007 May;58(5):469-473.

Fagien S, Cox SE, Finn JC, Werschler WP, Kowalski JW. Patient-reported outcomes with botulinum toxin type A treatment of glabellar rhytids: a double-blind, randomized, placebo-controlled study. *Dermatol. Surg.*, 2007;33(1 Spec No):S2-S9.

Sommer B, Zschocke I, Bergfeld D, Sattler G, Augustin M. Satisfaction of patients after treatment with botulinum toxin for dynamic facial lines. *Dermatol. Surg.*, 2003 May;29(5):456-460.

US Food and Drug Administration. FDA Requires Boxed Warning for All Botulinum Toxin Products. http://www.fda.gov/NewsEvents/Newsroom/PressAnnouncements /ucm149574.htm.

Wenzel R, Jones D, Borrego JA. Comparing two botulinum toxin type A formulations using manufacturers' product summaries. *J. Clin. Pharm. Ther.*, 2007; 32(4):387-402.

Monheit G, Carruthers A, Brandt F, Rand R. A randomized, double-blind, placebo-controlled study of botulinum toxin type A for the treatment of glabellar lines: Determination of optimal dose. *Dermatol. Surg.*, 2007;33(1 Spec No.):S51-S59.

Ascher B, Zakine B, Kestemont P, Baspeyras M, Bougara A, Santini J. A multicenter, randomized, double-blind, placebo-controlled study of efficacy and safety of 3 doses of botulinum toxin A in the treatment of glabellar lines. *J. Am. Acad. Dermatol.*, 2004;51(2):223-233.

Rzany B, Ascher B, Fratila A, Monheit GD, Talarico S, Sterry W. Efficacy and safety of 3- and 5-injection patterns (30 and 50 U) of botulinum toxin A (Dysport) for the treatment of wrinkles in the glabella and the central forehead region. *Arch. Dermatol.*, 2006;142(3):320-326.

Ascher B, Zakine B, Kestemont P, et al. Botulinum toxin A in the treatment of glabellar lines: scheduling the next injection. *Aesth. Surg. J.*, 2005;25:365-375.

Brandt F, Swanson N, Baumann L, Huber B. Randomized, placebo-controlled study of a new botulinum toxin type a for treatment of glabellar lines: efficacy and safety. *Dermatol. Surg.*, 2009;35(12):1893-1901.

Rosales RL, Bigalke H, Dressler D. Pharmacology of botulinum toxin: Differences between type A preparations. *Eur. J. Neurol.*, 2006;13(Suppl 1):2-10.

Karsai S, Adrian R, Hammes S, et al. A randomized, double-blind study of the effect of Botox [R] and Dysport [R]/Reloxin [R] on forehead wrinkles and electromyographic activity. *Arch. Derm.*, 2007;143:1147-1149.

Wiest LG. An overview of the cosmetic treatment of facial muscles with a new botulinum toxin. *Acta. Dermatovenerol. Croat.*, 2009;17(1):48-53.

Xeomin [R]. Summary of Product Characteristics (Germany), *Merz. Pharmaceuticals,* 2009.

Bocouture [R]. Summary of Product Characteristics (Germany), *Merz. Pharmaceuticals,* 2009.

Stenmark P, Dupuy J, Imamura A, et al. Crystal structure of botulinum neurotoxin type A in complex with the cell surface co-receptor GT1b: insight into the toxin-neuron interaction. *PLoS Pathog.*, 2008;4(8):e 1000129.

Eisele K-H, Taylor HV. Dissociation of the 900 kDa neurotoxin complex from C. botulinum under physiological conditions. *Toxicon.*, 2008;51(Suppl 1):S10.

Eisele KH, Grein S, Mander GJ, Taylor HV. Role of complexing proteins in pharmaceutical botulinum neurotoxin formulations. Poster #733, presented at IMCAS. Paris, France, 2009.

Dressler D. Pharmacological aspects of therapeutic botulinum toxin preparations. *Nervenarzt,* 2006;77:912-921.

Roggenkamper P, Jost WH, Bihari K, et al. Efficacy and safety of a new Botulinum Toxin Type A free of complexing proteins in the treatment of blepharospasm. *J. Neural. Transm.*, 2006;113(3):303-312.

Baumann LS, Halem ML. Botulinum toxin-B and the management of hyperhidrosis. *Clin. Dermatol.*, 2004;22(1):60-65.

Jacob CI. Botulinum neurotoxin type B--a rapid wrinkle reducer. *Semin. Cutan. Med. Surg.*, 2003;22(2):131-135.

U.S. Food and Drugs Administration. FDA Gives Update on Botulinum Toxin Safety Warnings; Established Names of Drugs Changed [online]. Available on URL: http://www.fda.gov/NewsEvents/Newsroom/ PressAnnouncements/ucm175013.htm .

Flynn TC. Myobloc. *Dermatol. Clin.*, 2004 Apr;22(2):207-211.

Sadick NS, Faacs. Botulinum toxin type B for glabellar wrinkles: a prospective open-label response study. *Dermatol. Surg.*, 2002;28(9):817-821.

Sadick NS. Prospective open-label study of botulinum toxin type B (Myobloc) at doses of 2,400 and 3,000 U for the treatment of glabellar wrinkles. *Dermatol. Surg.*, 2003;29(5):501-507.

Alster TS, Lupton JR. Botulinum toxin type B for dynamic glabellar rhytides refractory to botulinum toxin type A. *Dermatol. Surg.*, 2003;29(5):516-518.

Jacob CI. Botulinum neurotoxin type B-a rapid wrinkle reducer. *Semin. Cutan. Med. Surg.,* 2003;22(2):131-135.

Spencer JM, Gordon M, Goldberg DJ. Botulinum B treatment of the glabellar and frontalis regions: a dose response analysis. *J. Cosmet. Laser. Ther.,* 2002;4(1):19-23.

Flynn TC, Clark RE. Botulinum toxin type b (MYOBLOC) versus botulinum type a (BOTOX) frontalis study; rate of onset and radius of diffusion. *Dermatol. Surg.,* 2003;29(5):519-522.

Clinical Trials website. Available at http://www.clinicaltrials.gov/ct2/show /NCT00752297?term=purtoxandrank=3.

Stone AV, Ma J, Whitlock PW, Koman LA. Effects of Botox [R] and Neuronox [R] on muscle force generation in mice. *J. Orthop. Res.,* 2007;25:1658-1664.

Medy-Tox website. Available at www.medy-tox.co.kr.

Hunt T, Clarke K. Potency of the botulinum toxin product CNBTX-A significantly exceeds labeled units in standard potency test. *J. Am. Acad. Dermatol.,* 58(3):517-518 (2008 Mar).

Pierard GE, Lapier CM. The microanatomical basis of facial frown lines. *Arch. Dermatol.,* 1989;125:1090-1092.

Kahn DM, Shaw RB. Overview of current thoughts on facial volume and aging. *Facial. Plast. Surg.,* 2010 Oct;26(5):350-355.

Charles Finn J, Cox SE, Earl ML. Social implications of hyperfunctional facial lines. *Dermatol. Surg.,* 2003 May;29(5):450-455.

Mac Pherson S. Self-esteem and cosmetic enhancement. *Plast. Surg. Nurs.,* 2005;25(1):5-20.

Leffell DJ. Foreword. In: Kaminer MS, Dover JS, Arndt KA, editors. Atlas of Cosmetic Surgery. 1st ed. Philadelphia, Pa: W. B. Saunders Company; 2001. p. ix–x.

Fagien S. Botox for the treatment of dynamic and hyperkinetic facial lines and furrows: adjunctive use in facial aesthetic surgery. *Plast. Reconstr. Surg.,* 1999;103:701-713.

Carruthers J, Carruthers A. The evolution of botulinum neurotoxin type A for cosmetic applications. *J. Cosmet. Laser. Ther.,* 2007 Sep;9(3):186-192.

In: Botulinum Toxin
Editors: Fausto E. Marino and Sebastiano De Luca
ISBN: 978-1-62257-185-7

Chapter IV

The Use of Botulinum Toxin in the Treatment of Pain

[1]Dominique Batifol and [2]Pierre-Jacques Finiels
[1]Department of Oro-facial Surgery, Montpellier University Hospital, Montpellier, France
[2]Department of Neurosurgery and Spine Pathology, Kennedy Medical Center, Nîmes, France

Abstract

The widespread use of Botulinum toxin (BT) is nowadays well accepted in the treatment of pain of spastic origin and also for aesthetic purposes. In addition, a more accurate knowledge of its action has lead to new methods of research which seem to show new indications, many of them totally unknown, even in the specialized medical community.

The aim of this work is to present our experience and first results in such cases of non-conventional indications.

1. Usefulness of Botulinum Toxin in the Treatment of Algid Dysfunction of the Temporo-Mandibular Joint

A prospective study has begun in our department in 2007, following the work on the knee joint published by Mahowald et al. in 2006 [1].

Without the real possibilities to make a study versus placebo, such as washing of the joint with buffered saline is known to have a therapeutic effect [2], we have chosen to correlate results of treatment usually tried in this form of pathology, excluding surgery, in all of the patients we have had to treat. Treatments first attempted included physiotherapy, usual grade I to III analgesic medications, dental groove, intra-muscular injections of botulinium toxin (either in masseter or temporal muscle) and intra-articular injections of hyaluronic acid.

When all of those attempts led to no improvement, i.e. visual analogical scale (VAS) evaluation of the pain still greater than 5/10, then the purpose of intra-articular injections of botulinum toxin was made.

Release or diminution of the pain was the only goal of the injection, without any hope of increasing mouth aperture or diminishing crackles in the joint.

Patients were injected with 30U Botox* (Allergan Pharmaceuticals, Westport, Ireland) in each articulation, injections never being systematically done bilaterally.

The choice of using 30U was made after some trials using 20U alone before the beginning of the study, which had quickly proved to be insufficient.

Technique: injections are made in the classic aseptic way under local anesthesia. The dilution is 100U Botox* in 1ml of 0, 9% sodium chloride in a graduated syringe. Without the need to use fluoroscopy, anatomical landmarks are used as follows: posterior aspect of ascending branch of the mandible towards the condyle, inferior border of the zygomatic arch and, as possible, perception of the movement of the condyle at mouth's aperture (Figure 1a) Injection is done under aperture of the mouth, in the posterior part of the joint (Figure 1b). Aperture measurement is made before and after injection allowing certainty of being made in the joint, as washing effect described above leads towards brief amelioration of pain and aperture capacity. Self-evaluation questionnaires about quality of life and evaluation of pain are given to patients, who are systematically reviewed one month after injection.

Results: as usual, effect rises gradually; beginning 15 days after and reaching its maximum 1 month after injection. Effect still remains for 4 to 5 months in our experience, with a definitive release in some patients. To date, this technique was tested in 37 patients, with results as follows:

- VAS: 0/10 at 1 month in 17 (46%) patients, some of them having a VAS=10/10 before injection;
- VAS between 2 and 4/10, intermittent in presentation, in 11 (30%) patients;
- 9 patients (24%) showing no quantifiable amelioration.

Reasons for failure can be understood as follows:

- The product used: a very small part of the general population is known as being not responsive to botulinum toxin therapy. Even taking this into consideration, the real number cannot explain the attrition rate;
- The physician making the injection: even with much experience, realization of an injection without the help of fluoroscopy remains challenging in such rearranged and very narrow joints. However, post-injection transient amelioration of mouth aperture and its measurement leads to think that such mistakes are quite rare;
- Operating procedure: quantity chosen for injection is not made at random but conforms to the results of previous trials. Protocol is not perfect, has already been changed and could be again in the future;

- The patient: those being potentially chosen for this kind of treatment are generally seen for months before injections, trying all alternatives generally for 6 to 9 months before reaching status authorizing protocol inclusion. No psychological evaluation is made in the way of simplify inclusions.

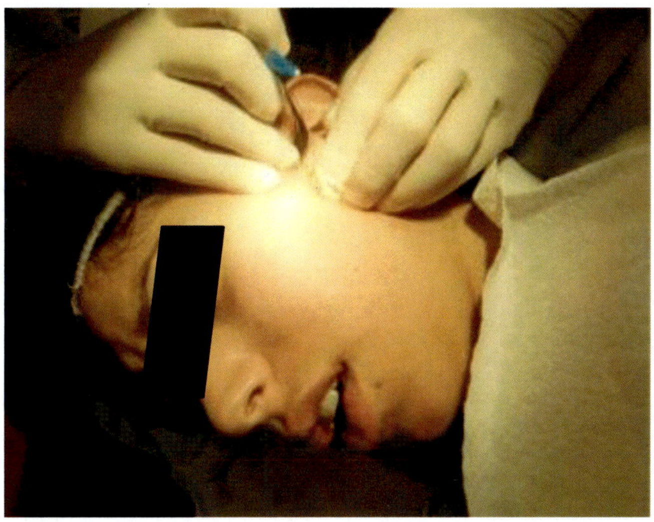

Figure 1a. Location of anatomic landmarks.

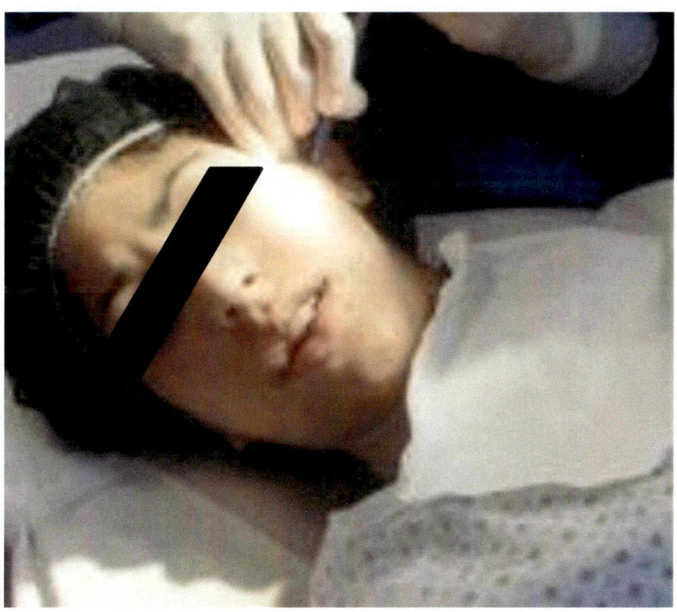

Figure 1b. Injection in the temporo-mandibular joint.

Nevertheless, first results seem to be encouraging, the number of "non-responders" diminishes as the number of cases treated increases.

Mechanism of action of BT on pain is known and well studied since a few years. It seems that BT would have some neuro-inhibiting effect, blocking transmission of pain information.

Inhibition of peripheral nociceptive receptors, by preventing neurotransmitters (as P-substance or glutamate) secretion, could lead to a real desensitization either on peripheral nerve endings or at the center level [3].

BT is well tolerated and has a good level of acceptance in such populations, for reasonable costs and with a quite simple and easy to begin technique. The action seems to be only analgesic and ought to be confirmed by a multi-center randomized study.

2. Use of BT in Irradiated Patients for the Treatment of Cancerous Disease of the Oro-facial Region

In this very particular case, patients come with retraction problems [4] or severe trismus [5]. Tissue retraction inhibits feeding, phonation and could change the facial aspect. Spasms (peripheral nerves dysfunction due to ischemia or neurons myelinoclasis), neuropathy and fibrosis (cytokines and growth factors release) lead to important pain [6]. Those adverse effects remain even after the end of radiotherapy treatment and are usually badly supported. They are not infrequent and very difficult to treat. BT seems to act as a muscle relaxant by diminishing radio-induced muscular spasms and inducing direct analgesic effect [7].

Nowadays, about 50% of cancers benefit from radiotherapy, either as principal care modality or second line therapy if not only for analgesic attempt. Irradiating tumorous cells is far from being without side effects on healthy tissues and bordering organs [8]. In the oral and facial region, after a long time, tissular fibrosis with important trismus, skin retractions, spasms and cramps appear, associated with important and finally debilitating pain [9]. This kind of pain, of neuropathic nature, has been shown to be accessible to BT use [10, 11].

To date, 17 patients (between 54 and 77 years of age at the first contact, the first injection coming 6 months to 5 years after radiotherapy) have been recruited in our protocol for treatment of this kind of condition: 9 having musculo-cutaneous retractions with cervical adhesions and muscular fibrosis, 6 a severe trismus and 2 a facial deformity, especially found around mouth aperture.

Technique: Injections are made every 3 months, in the muscles presenting post-radic modifications. Doses used should be important as the action of the toxin seems to be slowed down by fibrosis, acting against diffusion and so justifying multiplication of injection points and an increase of the doses commonly used, as far as 300 to 400U Botox* in the same session. Gains are in mm of oral aperture in conjunction with physiotherapy, and the sooner the treatment is started the better the result seems to be. The BT can be used as well in the cervical musculature (Figure 2a), even in patients having dysphagia before beginning of the treatment. In case of trismus, targeted muscles are masseters and temporal muscles, and sternocleidomastoidmuscle and anterior part of the biventer cervicis, in case of skin retractions at the cervical level. In case of important deformation of the face at the irradiated side with fibrosis beginning, leading to facial palsy's aspect, with deformation of the mouth and disclosure of the lips, the zygomaticus major and minor, as well as the alveololabialis and levator anguli oris muscles are injected, doses used ranging from 100 to 150U Botox*.

Injections are made with a dilution of 100U Botox* for 1ml of normal saline, under electromyography control.

Results: Amelioration could be seen after 2 to 3 injections, with pain diminution, widening of buccal aperture and remobilization of the cervical region in case of skin retractions (Figure 2b). To date, our best result was on buccal aperture coming from 0 to 28mm in 18 months; another patient (the last one to be recruited) showed amelioration in the same way from 3 to 6 mm, after only one injection.

Results are globally satisfactory for the patients, either as ameliorating function (better buccal aperture, allowing normal feeding or normalization of cervical stature, eliminating pain), or purely cosmetic in case of oral and facial deformation, with good outcome on patient confidence.

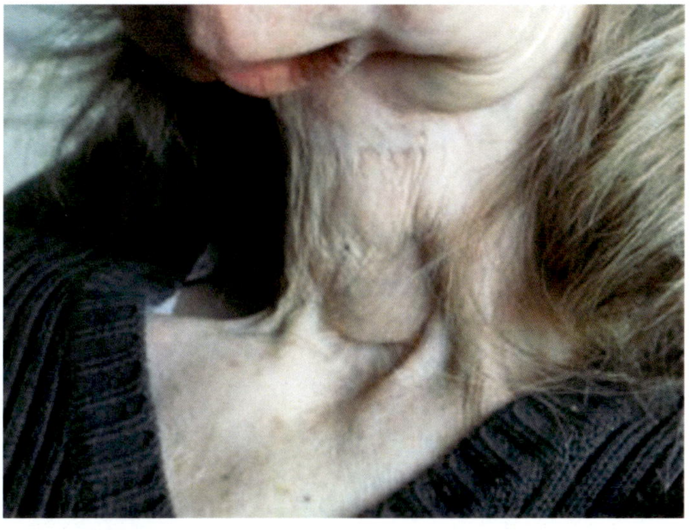

Figure 2a. Example of tissular retraction following radiotherapy.

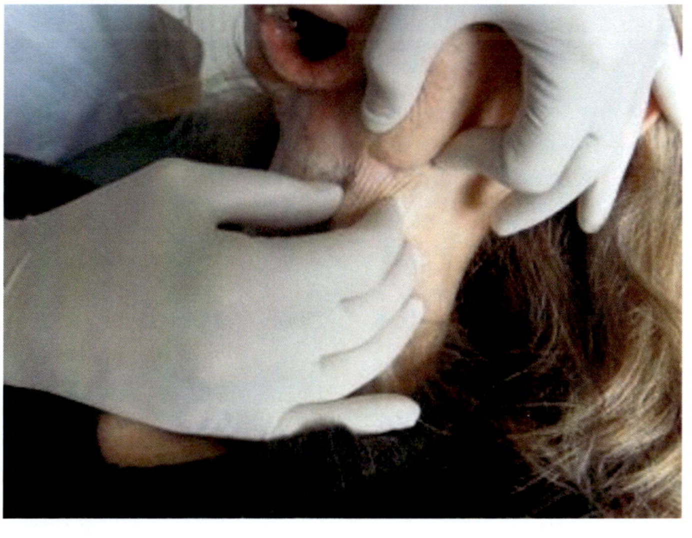

Figure 2b. Same patient, after 3 injections.

Only one case of failure needs to be noted, in a patient treated 5 years earlier and presenting a settled fibrosis. BT does not have an action on fibrosis, but only on residual muscles, showing the importance of beginning the treatment as soon as possible after the end of radiotherapy. It is important to state that inpatients benefit from physiotherapy as well, acting in synergy with BT.

To date, no adverse effects have been found in the follow up of our patients. Doses used could seem to be important, long-term and regular treatment being mandatory, showing a proven amelioration in quality of life.

3. Usefulness of Botulinum Toxin Injections in the Treatment of Post-Operative Pain after Cervical Spine Surgery

Neck and shoulder pain (axial symptoms) after anterior inter-body fusion and/or laminoplasty in patients with cervical spondylotic myelopathy is a common concern in a neurosurgical daily practice of which the prevalence has been seldom studied [12]. Its pathogenesis is certainly not of single origin. Disabling for the daily living conditions of patients, their treatment, using traditional pharmacological drugs and physical rehabilitation is often unsuccessful. Paying particular attention to results of the use of botulinum toxin in cervical dystonia and spasmodic torticolis [13, 14, 15], a study was designed to attempt to show efficacy of its use in such a condition.

Technique: It has been published elsewhere [16, 17]. Briefly, muscles concerned by injection were in frequency regression: trapezius, supra spinalis, splenius capitis and rhomboids. Injections were made using 20 to 100U Botox* without going beyond 300U once in the same patient, performed every 3 months if necessary (Figure 3]. The conservative treatment consisted of course of thiocolchicosid (16mg/day) and physical rehabilitation. Lordosis angle was calculated on lateral sitting radiographs in neutral position immediately post-operatively and 1.5 months after injection and correlated to amelioration in pain evaluated by the VAS.

Results: 215 patients, who have been operated on in the same institution for cervical spondylotic myelopathy were subsequently followed-up; 38 among them presented post-operative neck pain and were enrolled differently either in a course of botulinum toxin injections for 19 of them and conservative treatment for the remaining 19 patients. No visible amelioration was found on X-rays in 3 patients after BT use and in 11 after conservative treatment. In 16 cases, after an average of 3 injections, gain in lordosis averaged 11.3° and VAS score was reduced by 4.6 points, as compared to 4.7° and a reduction of 0.6 points after conservative treatment. An example could be seen in the following pictures (Figure 4):

Prevalence of post-operative neck pain after anterior cervical discectomy with inter-body fusion has been estimated to reach 19% of cases (8.6% in the present series) with possibility of increasing up to 60% after laminoplasty (14.3% in our series) [12], even if a more recent study has shown, conversely, a global satisfaction of patients after cervical surgery in 93% of cases [18], in a prospective study in 87 cases, surgery being made in these cases for chronic cervical pain. Exact causes of post-operative pain in such conditions are not perfectly understood, even though muscular and ligamentous components would seem to be acquired.

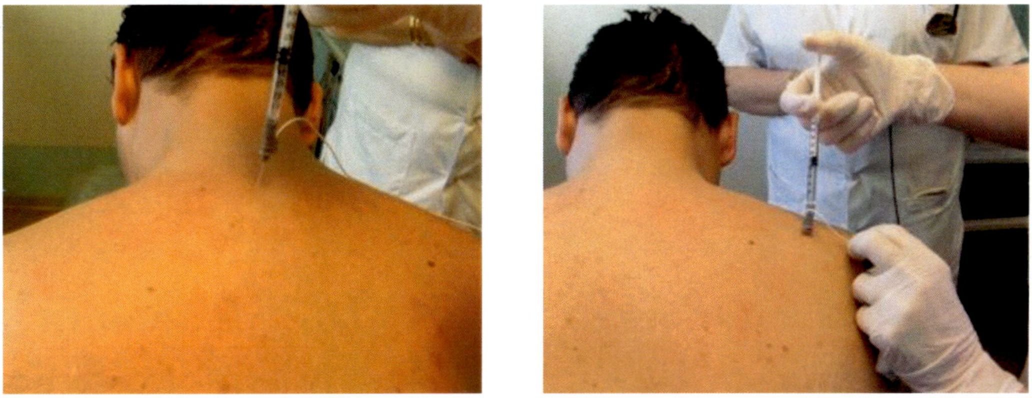

Figure 3. Examples of injection in trapezius and supra-spinalis muscle.

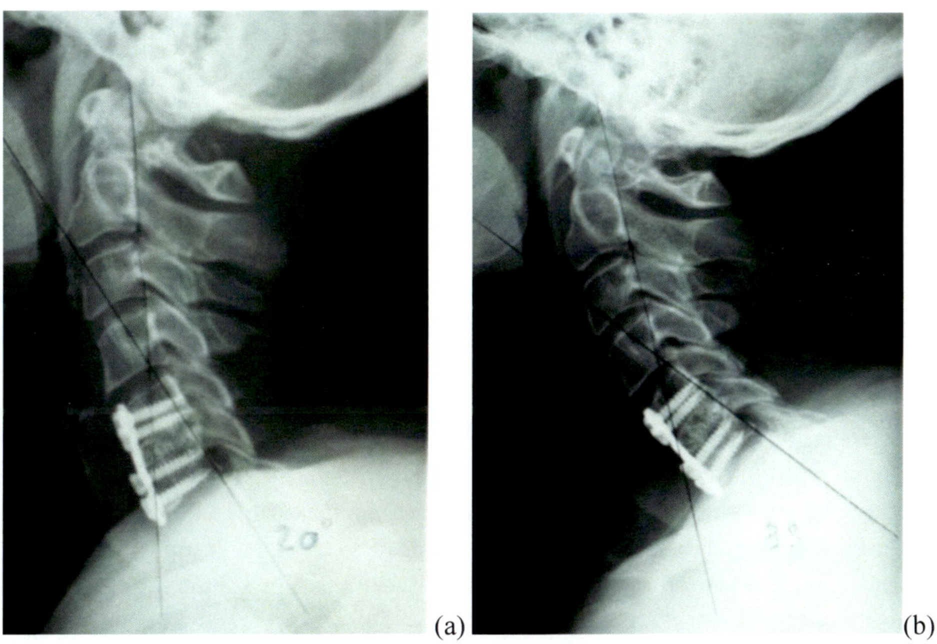

Figure 4. Lateral sitting radiographs immediately post-operative (1a) and 2 month (1b) after the last injection, with gain in lordosis calculated as 19°, in 50 years old female patient operated upon with laminectomy and C5-C6 arthrodesis. VAS amelioration was of 5 points.

In our experience, using the type A toxin, the mean dose injected, depending on the muscle volume, being around 50 (20-100) U Botox* by muscle injected. Intraoperative electromyography recording seems to be imperative, to confirm accurate localization of the tip of the needle in the center of the muscle belly. First signs of efficacy (diminution of sensations of muscular contractions in the groups treated) appear after ten days and exhaust after 3 months, with the need of new injections, as reported by others in lateral indications [19, 20]. Renewing of injections should be made every three months, avoiding less delay to reverse possibility of apparition of anti-toxinantibodies, which could cancel efficacy. 16 among 19 patients treated have shown real symptom amelioration, making a wider longitudinal investigation necessary to confirm this first impression. The mean cost of a 100 U Botox* injection is $220 in a hospital institution, compared to the cost of long course of classic pharmacological treatment and physiotherapy, and loss of production in case of a long period of sick leave. First results would seem to justify a multi-centric prospective study in the future.

4. Usefulness of BT in Keloid Treatment: Anti-Inflammatory Effect?

Keloids are found in predisposed patients [21] after skin trauma of any cause; wound, incision, burn, injection, inflammatory cutaneous lesion [22]. Scars can be thought as intra-dermal fibrous tumors, mostly found in "high-risk" areas that could be the upper dorsal region, near sternal, facial or round the ear locations [23].

Contrary to hypertrophic scars, keloids have no spontaneous trends to resolution.

We report here a single case of keloid which has been successfully treated in a young boy after surgery for protruding ears in 2008. This boy would seem to have no predisposed condition, apart for having had a hypertrophic scar after wound of the thigh, began to develop an auricle deformation one year after surgery, which slowly increased for one more year, even after diverse attempts of local treatment (corticosteroid injections essentially, radiotherapy having been proposed but not finally used). Re-intervention was declined by the surgeons consulted upon. By reference to the work of Zhibo and Miaobo [24], an attempt of using BT in such a condition was suggested and accepted after the patient, minor at this time, and his mother had been fully informed. In November 2010, at the first consultation, presentation was with 2 voluminous scars, especially the right one, being reddish and warm of aspect, and painful with important local inflammatory reaction (Figure 5a).

Technique: Only 20U Botox* were injected in the right scar, which seemed to be the most painful and voluminous. It was advocated to make this choice as a "test" to point answering to BT, authorizing future adaptation of the treatment, and making it acceptable without premedication or local anesthesia, as injections in a keloid reveal to be particularly painful.

Results: 3 days after injection, bleeding from the auricle was indicated by the mother, coming from the scar. In fact, it consisted of a serous rosé liquid, which, after release, left behind a seriously diminished volume of the ear, without pain but normalized color and temperature. Scar tissues were soft at palpation. The dose was notably insufficient to lead to 3 months result between 2 injections, as expected, and a new injection of corticosteroid was mandatory.

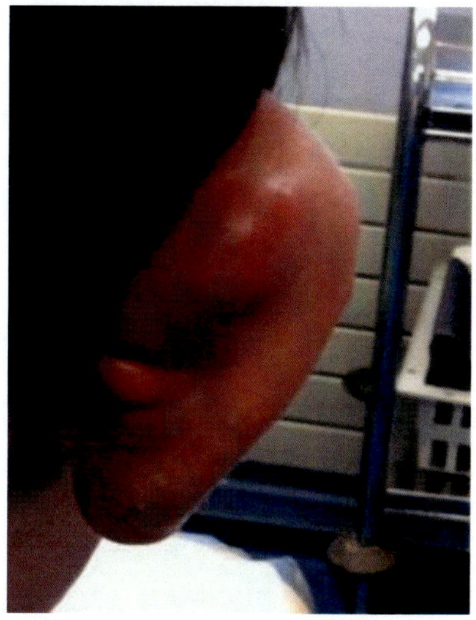

Figure 5a. Local aspect of the right auricle.

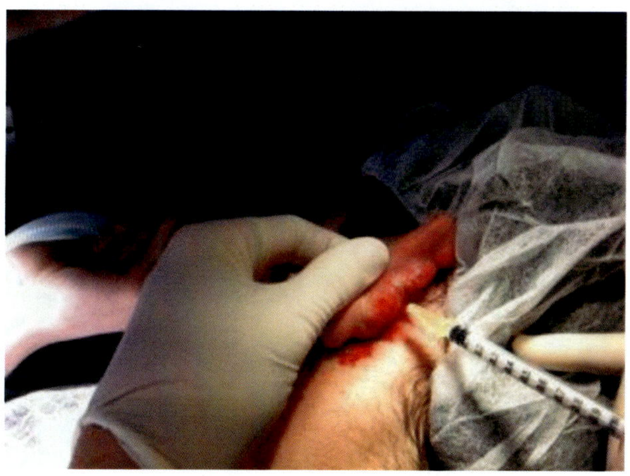

Figure 5b. Same patient, at the moment of second injection before treatment.

Figure 6. Same patient as above, at the end of the treatment, showing effect after surgery on the right auricle. On the left side, only BT injections were made.

The second injection (Figure 5b), in February 2011, was of 70U Botox*: 50U in the right ear and 20U in the left one. This time, pre-medication was done by 100mg hydoxyzine and under open anesthesia with a 50% mixture of oxygen and nitrous oxide. Great amelioration of the clinical presentation was noted, with resolution of the inflammatory process, leaving notably a voluminous scar, which could only be resolved by surgery. After new advice by the initial surgeon, re-intervention was decided upon for the right ear. New injections were made, rising up to 100U Botox* in the right ear and 30U in the left one on May 18th, preceding re-intervention which was conducted on June 16th. The patient was allowed to go home 4 days later with a bilateral compression device. Since this time, 3 other injections were made on August 17th, November 9th and February 22th, 2012, with the same protocol as the first two, and 70U Botox* to the right and 60U to the left for the last injection. Results are satisfactory, either in the right side at 8 months after surgery, or in the left one, without surgery. 3 more injections, leaving 3 months between each, are programmed, before spacing them out under close supervision of any local inflammatory signs (Figure 6).

This case report point out the anti-inflammatory effect of type A BT, as reported by a few teams, especially in mice [26], by inflammation mediators (as glutamate or P-substance) liberation inhibition. It could be this mechanism which is involved here, but a second one was advocated by others [27]: a possible effect on local tensions and cellular proliferation inhibition. If BT local effect on muscular tensions is well-known during the scarring process [28], the effect on hypertrophic scarring by the reduction of cell cycle distribution of fibroblasts leading to it seems to be a new acquisition which has to be confirmed.

5. BT Use Trials on Specific Neuralgias

Trigeminalal neuralgia is a debilitating condition, with somewhat bad tolerance and adverse effect either of medications devoted to its treatment, or surgical ones, leading to definitive after-effects. Knowing about what has been published on neuropathic pain [29], we have begun to drive a prospective randomized study in our institution, with subcutaneous administration of BT type A in the treatment of such condition. Injections are made either on the whole surface of cutaneous projection of the nerve [30] or at the level of the trigger zone wherever it could be localized, i.e. cutaneous or gingival [31], the choice of the way of it being done at random. The aim of the study is to show if a real efficacy does exist in this kind of pathology and also if there is an interest of only targeting trigger zone, which could, in the end, lead to low levels of toxin use.

Technique: injection are made with dilution of 100U Botox* for 1ml of sodium chloride, in a subcutaneous way (Figure 7), without using electrophysiological recording. Doses used range from 15 to 50U Botox*, 2 to 3U being used at every puncture point, after having prepared the skin with local anesthetic cream (lidocain - prilocain) 1 hour before. Self-evaluations questionnaires, using VAS, are given to patients, allowing follow-up of analgesics use.

Results: to date, the study is still in process, and has not been published. For the moment, we have had 16 patients totally freed of pain, 10 after injections of cutaneous projection zones and 6 after injections of the trigger zone only. Amelioration begins after 1 to 2 weeks, being sometimes variable among patients, leading to modifications of doses used. In 2 cases, the result remains after 6 months and no other injection has been done; in 3 cases, light facial asymmetry has been noted, and afterward corrected by contra-lateral injections, but has not driven to relinquishment from the patients. No significant differences has been detected after both techniques of injections, only one patient having a well individualized trigger zone with painful irradiations, necessitating occasional use of analgesics after partial amelioration, leading to modification of pain perception. Results come from patients quoting themselves their pain 0/10 on VAS without the use of analgesics to 2/10 with lowing the use of carbamazepin at 400mg a day. These results need to be confirmed by further completion of the study.

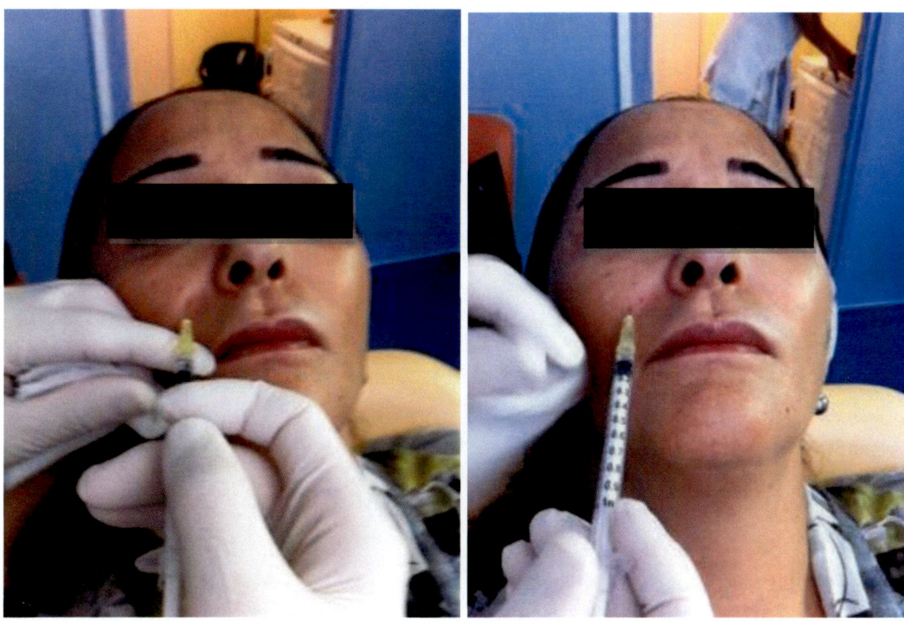

Figure 7. Example of subcutaneous injections in a case of right trigeminal neuralgia.

Occipital (so called Arnold's) neuralgia comes in two thirds of the cases with unilateral presentation, bilateral in the other third [32]. Already available therapeutics are divided between analgesic and antiepileptic medications, direct ablative surgery (neurotomy) and thermal radiofrequency treatment (thermocoagulation) [33]. Analgesics are seldom effective, and antiepileptic medications such as carbamazepin, gabapentin or pregabalin are far from being without adverse effects. Thermal radiofrequency treatment should be done by a trained surgeon regarding the risk of definitive neural damage [33, 34), and allows pain release in about three quarter of the cases.

Since 2011, a randomized prospective study has begun in our institution, in order to compare results of patients receiving BT injections versus those benefiting from radiofrequency treatment. To date, 23 patients have been injected and 15 others has been treated using thermal radiofrequency lesion.

Technique: doses are ranged from 100 to 150U Botox* in double dilution, i.e. 100U Botox* in 2ml sodium chloride. Injections are made on a prepared skin with local analgesic cream applied 45min before, and sterilized. A topping of the zone to be treated is made by multiple subcutaneous injections of 2 to 3U at each point of puncture (Figure 8). Injections begin at the visible probe of the nerve, in the so-called Tillaux's triangle, thus are made in the area of pain irradiation, i.e. occipital, parietal and even on supra auricle and near medial vertex regions.

Results: all patients were received with VAS score between 7 and 10/10 before injection. After injection, 22 patients have had a notable amelioration, with a remaining VAS score at 0 in 13 (56%), and between 2 and 5/10 in 9(40%) cases with remaining occasional dolorous crises, coming with less frequency. In one case, no effect was seen. In comparison, patients benefiting from radio frequency treatment were freed from pain in 10 cases (67%), ameliorated but necessitated to continue pain medication in 2 cases (13, 5%) and with no amelioration at all in 3 cases (19, 5%).

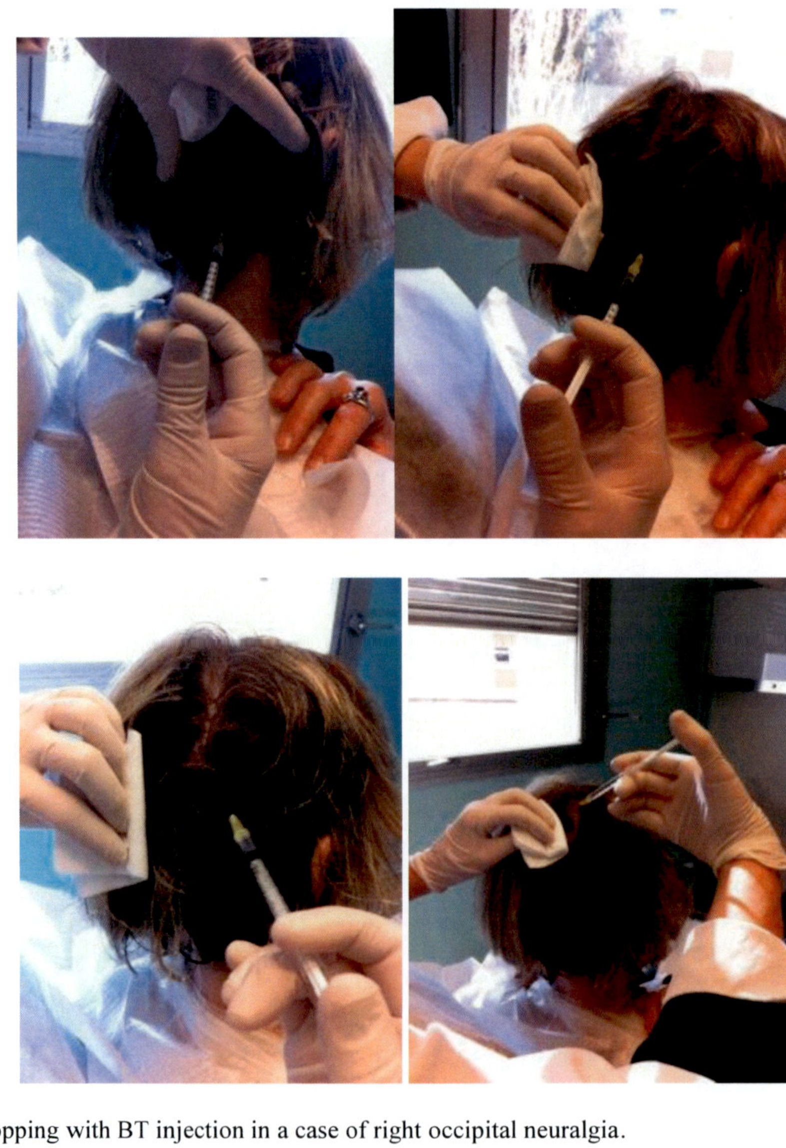

Figure 8. Topping with BT injection in a case of right occipital neuralgia.

For the patients with clear amelioration, the same doses were used for new treatments, and increased with a little more success in the others. No new attempt was made for the patient in whom treatment has failed.

Conclusions: to the best of our knowledge, this is the first time that the way of searching inside use of botulinum toxin for the treatment of painful pathologies is drawn in the literature. A lot of knowledge and specific fundamental research seems mandatory for understanding a real way of action in each type of pain as much as multi center larger cohort investigation to confirm first results.

Disclosure

No help or financial interests were obtained for this work.

Acknowledgments

The authors would like to thank their interns, Marie de Boutray and Sophie Galmiche for bibliographic research assistance.

References

[103] Mahowald, M.L., Singh, J.A., Dennis Dykstra, D. Long term effects of intra-articular botulinum toxin A for refractory joint pain. *Neurotox Res* 2006;9 (2-3):179-188.

[104] Kawai N., Tanaka E., Miyauchi, M. et al. Influence of additive hyaluronic acid on the lubricating ability in the temporo-mandibular joint. *J Biomed Matter Res A* 2004; 70 (1): 149-153.

[105] Cui, M., Khanijou, S., Rubino, J., Aoki, R. Subcutaneous administration of botulinum toxin A reduces formalin-induced pain. *Pain* 2004; 107: 125-133.

[106] Stubblefield, M.D., Levine, A., Custotio, C., Fitzpatrick, Th. The Role of Botulinum Toxin Type A in the Radiation Fibrosis Syndrome: A Preliminary Report. *Arch. Physical Med. Rehabilit.* 2008; 89 (3): 417-421.

[107] Wang, C.J., Huang, E.Y.,Hsu, H.C. et al. The degree and time-course assessment of radiation-induced trismus occurring after radiotherapy for nasopharyngeal cancer. *Laryngoscope* 2005;115:1458-60.

[108] Burton, A.W. et al. Chronic pain in the cancer survivor: a new frontier. *Pain Med* 2007; 8:189-98.

[109] Hobson, D.E., Kerr, P., Hobson, S. Successful use of botulinum toxin for post-irradiation unilateral jaw neuro-myotonia. *Parkinsonism & Related disorders,* 2009, 15 (8): 617-618.

[110] Kulik, C., Mazurier, J., Lartigau, E. Probabilités de contrôle tumoral et de complications (TCP/NTCP) après radiothérapie : aspects méthodologiques, physiques et biologiques. *Cancer/Radiother* 2002; 6 (1): 155s-165s.

[111] Greenfield, M.M., Stark, G.M. Post-irradiation neuropathy. *Am J Roentgenol* 1948; 60:617-22.

[112] Aoki, K.R. Review of a proposed mechanism for the anti-nociceptive action of botulinum toxin type A. *Neurotoxicology* 2005; 26: 785-93.

[113] Cui, M., Li, Z., You, S., Khanijou, S., Aoki, K.R. Mechanisms of the anti-nociceptive effect of subcutaneous BOTOX: inhibition of peripheral and central nociceptive processing. *Naunyn Schmiedebergs Arch Pharmacol* 2002; 365:R17.

[114] Hosono, N., Yonenobu, K., Ono, K. Neck and shoulder pain after laminoplasty. A noticeable complication. *Spine* 1996; 21 (17): 1969-1973.

[115] Boghen, D., Flanders, M. Effectiveness of botulinum toxin in the treatment of spasmodic torticolis. *Eur Neurol* 1993; 33 (3): 199-203.

[116] Poewe, W., Schelosky, L., Kleedorfer, B., Heinen, F., Wagner, M., Deuschl, G. Treatment of spasmodic torticolis with local injections of botulinum toxin. One-year follow-up in 37 patients. *J Neurol.* 1992; 239 (1): 21-25.

[117] Swope, D., Barbano, R. Treatment recommendations and practical applications of botulinum toxin treatment of cervical dystonia. *Neurol Clin* 2008; 26 (Suppl 1): 54-65.

[118] Finiels, P.J., Batifol, D. Utilisation de la toxine botulinique dans les cervicalgies post opératoires en chirurgie rachidienne ; Résultats préliminaires. *Neurochirurgie:* 2010,56:374-381.

[119] Finiels, P.J., Batifol, D. Usefulness of botulinum toxin injections in the treatment of post-operative pain after cervical spine surgery. *Pan Arab J Neurosurg* (Under Press)

[120] Garvey, T.A., Transfeldt, E.E., Malcom, J.R., Kos, P. Outcome of anterior cervical disectomy and fusion as perceived by patients treated for dominant axial-mechanical cervical spine pain. *Spine* 2002; 27 (17): 1887-1895.

[121] Arnaud, S., Batifol, D., Goudot, P., Yachouh, J. Non-surgical managment of traumatic injuries of the parotid gland and duct using type A botulinum toxin. *Plast Reconstr Surg* 2006; 117 (7): 2426-2430.

[122] Ellis, M., Laskawi, R., Rohrbach-Volland, S., Arglebe, C. Up-to-date report of botulinum toxin therapy in patients with drooling caused by different etiologies. *J Oral Maxillofac Surg* 2003; 61: 454-457.

[123] English. R.S., Shenefelt,, P.D. Keloids and hypertrophic scars. *Dermatol Surg* 1999;25:631-638.

[124] Rook, A., Wilkinson, D.S., Ebling, F.J., Champion, R.H., Burton, J.L. Hypertrophic scars and keloids. In: *Textbook of Dermatology,* 4th Edition. Blackwell Scientific Publication, Oxford, 1986;1832-1833.

[125] Crockett, D.J. Regional keloid susceptibility. *Br J Plast Surg* 1964;17:245-53.

[126] Zhibo, X., Miaobo, Z. Intralesional botulinum toxin type A injection as a new treatment measure for keloïds. *Plast Reconstr Surg* 2009; 124(5): 275e-277e.

[127] Cui, M., Khanijou, S., Rubino, J., Aoki, K.R. Subcutaneous administration of botulinum toxin A reduces formalin-induced pain. *Pain* 2004; 107: 125-133.

[128] Zhibo, X., Miaobo, Z. Botulinum toxin type A affects cell cycle distribution of fibroblasts derived from hypertrophic scar. *J Plast Reconstr Aesthet Surg* 2008; 61 : 1128-1129.

[129] Wilson, A.M. Use of botulinum toxin type A to prevent widening of facial scars. *Plast Reconstr Surg* 2006; 117(6): 1758-66.

[130] Luvisetto, S. et al. Anti-allodynic efficacy of botulinum neurotoxin A in a model of neuropathic pain. *Neuroscience* 2007; 145: 1-4.

[131] Bodoric, G.E. , Acquadro, M.A. The use of botulinum toxin for the treatment of chronic facial pain. *J Pain* 2002;3:21-27.

[132] Zuniga, C. et al. Beneficial effects of botulinum toxin type A in trigeminal neuralgia. *Arq Neuropsiquiatr.* 2008; 66 (3-A): 500-503.

[133] Bogduk, N. The anatomy of occipital neuralgia. *Clin. Exp. Neurol.* 1980; 17: 167-184.

[134] Privat, J.M., Finiels, P.J., Vlahovitch, B. *Névralgie occipitale : thermocoagulation du nerf d'Arnold* In : Céphalées et migraines, Masson, Paris, 1993 ; 204-209.

[135] Blume, H.G. Radiofrequency denaturation in occipital pain. A new approach in 14 cases. In: *Advances in pain research and therapy,* Springer, New York, 1976; 1: 691-698.

In: Botulinum Toxin
Editors: Fausto E. Marino and Sebastiano De Luca
ISBN: 978-1-62257-185-7

Chapter V

Lower Facial Contouring with Botulinum Toxin: Functional and Aesthetic Aspects

***Raffaele Rauso*[1,2,3] *and Roberto Amore*[4]**
[1]Centro Polispecialistico Santa Apollonia, S.M.C.V. (CE), Italy
[2]University of Foggia, Faculty of Dentistry, Italy
[3]University "La Sapienza", Department of Plastic Surgery, Rome, Italy
[4]Aesculapio Surgical Center, Pisa, Italy

Abstract

To achieve a slim and smooth contour line of the lower face, conventional treatment has concentrated on two approaches: surgical resection of the masseteric muscle or modelling ostectomy of the square-angled jaw. Modelling ostectomy of a square-angled jaw is mandatory when jaw's angles hyperostosis is present. When a square-angled face is due only to masseter muscle hypertrophy, our attention can be focused only on soft tissue remodelling. The resection of the masseter muscle is likely to cause a variety of side effects such as bleeding, hematoma, facial nerve injury, and asymmetric resection. Although the operation itself is relatively simple, it is difficult to predict the outcome.

Botulinum toxin injections into the masseter muscle can be used as an alternative, non invasive treatment, for the contouring of the lower face, however it is not associated with permanent results.

Masseteric hypertrophy is known as the asymptomatic enlargement of the masseter muscle and is commonly associated with abnormal habits such as nocturnal bruxism and daytime clenching.

The mechanism involved in the treatment of masseteric hypertrophy with botulinum toxin injection are somewhat different from those for treating facial hyperkinetic wrinkles. An intramuscular injection of botulinum toxin leads to temporary partial denervation at the area of the neuromuscular synapses and eventually causes muscle atrophy.

Occlusal parafunction, such as tooth grinding and tooth clenching, are unvoluntary mainly nocturnal habits that sometimes result in an hypertrophy of masseter muscle.

Otherwise, masseteric muscle hypertrophy aetiology seems to be correlated to congenital genesis, psycological factors, habits and occlusal disorders. Especially occlusal parafunction such as bruxism can be correlated with the genesis of masseteric muscle hypertrophy, because the more masticatory muscle are used the most their volume can grow-up. In clinical routinary practice occlusal parafunction are treated with occlusal devices or botulinum toxin injection.

Authors explain the role of botulinum toxin in the treatment of the square-angled face related to masseter muscle hypertrophy and the importance in treating occlusal disorders, if present, at the same time to prolong the contouring effect of the toxin injected.

Introduction

Nowadays, the use of botulinum toxin (BoNT) is well established in medical practice, not only for aesthetic purpose. The application of BoNT extends over many indications and medical specialties such as hemifacial spasm, strabismus and other ophthalmic disorders, bruxism, rhinitis, lacrimation, wrinkles, focal dystonias, foot dystonias, Tourette syndrome and other kinds of tics, hyperhidrosis, urologic disorders (hyperreflexive bladder, detrusor-sphincter dyssynergia, spasticity of the vescical sphincter, etc), gastrointestinal and proctologic disorders (achalasia, anal fissure, etc), and others. In several indications, the use of BoNT is well established and is considered the treatment of choice. The mode of approval varies from continent to continent, and from country to country. Another indication for the use of BoNT type A is the treatment of masseteric muscle hypertrophy (MMH) to achieve a lower facial contouring.

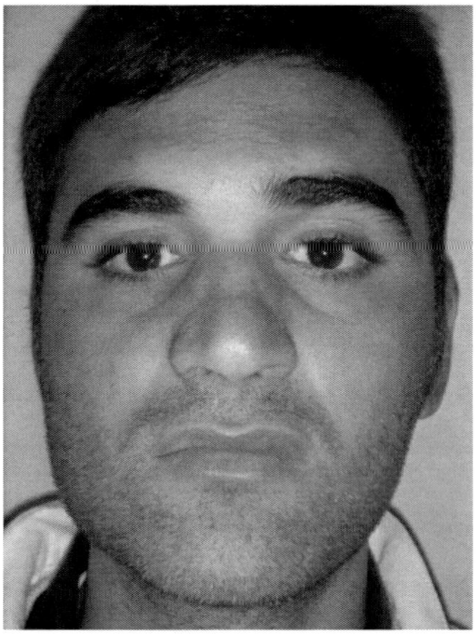

Figure 1. A patient affected by right unilateral MMH.

Figure 2. A patient affected by bilateral MMH.

The Masseteric Muscle Hypertrophy

Masseteric muscle hypertrophy is a clinical entity of unknown aetiology. It can be unilateral or bilateral and is commonly associated with abnormal habits such as nocturnal bruxism and daytime clenching. The hypertrophic muscle is also thought to cause secondary enlargement of the mandibular angle as a result of functional remodelling that occurs at the muscle insertion sites; in fact, sometimes mandibular angle hyperostosis can be associated to the MMH.

Clinically, the patient may present with a chronic painless firm enlargement of the angle of the jaw either unilaterally or bilaterally (Figure 1; 2). The enlargement is due to a combination of the hypertrophic muscle and enlargement of the underlying jaw bone. The highest incidence is in the second and third decades of life and there is no sex predilection.

The achievement of a slim and smooth contour line of the lower face has concentrated on two conventional approaches: surgical resection of the masseteric muscle or modeling ostectomy of the square-angled mandibule. In 1951, Converse described the resection of both the bone and the muscles, done at the same time, through an intraoral incision. Some years later, Whitaker also reported that the width of the lower face could be reduced through an ostectomy of cortical bone of the mandible, together with resection of the masseteric muscle. Baek found that the masseteric muscle volume is reduced over time as the muscle tone deteriorates when only the mandible was resected, whereas Hong et al. proved that after ostectomy of the mandible, the unused masseter muscle becomes atrophied over time. However, the resection of the masseter muscle is likely to cause a variety of side effects such as bleeding, hematoma, facial nerve injury, and asymmetry. Although the operation itself is relatively simple, it is difficult to predict the outcome. In this respect, even in the case of muscular hypertrophy, surgical treatment for the squared-angle jaw bone has been frequently applied. If muscular dystrophy is not as much as expected, however, the change in the lower face from the front is minimal. Thus, the need to reduce the lower facial muscular volume in a safe manner has been raised. In 2001, von Lindern et al. claimed that the masseter muscle might be reduced using botulinum toxin A (Dysport) in the case of masticatory hypertrophy.

Relationship between Occlusal Dental Pattern and Masseteric Muscle Hypertrophy

Tooth grinding and tooth clenching are involuntary, mainly nocturnal, habits that sometimes result in a hypertrophy of the masseter and temporalis muscles with an imbalance between the opening and closing muscle of the jaw. This leads to an alteration of mandibular condyle movement. Otherwise, the aetiology of the masseter hypertrophy seems to be related to congenital genesis, psychological factors, habits, and occlusal disorders. Occlusal parafunction, such as bruxism, can be especially correlated with the genesis of masseter hypertrophy, because the more the masticatory muscles are used, the most their volume increase. Usually, in clinical practice, occlusal parafunctions are treated with occlusal devices or botulinum toxin injection. Chikhani et al. in their study affirm that intramuscular injections of botulinum toxin in masseter and temporalis muscles equilibrate the balance between closing and opening muscles and treat masseteric hypertrophy, with improvement of face outline and the recovery of a normal cynesis of the temporomandibular joint. This could be considered an efficient treatment of bruxism and TMJ dysfunction, cheap with no lasting side effects. Similar findings are reported in the literature. Ahn et al. measured the change of maximum bite force after injection of Botulinum toxin type A (BoNTA) for the treatment of masseteric hypertrophy. The maximum bite force between the maxillary and mandibular first molar, before and after the BoNTA injection was measured; they concluded that the maximum bite force was significantly reduced after the injection, however it had gradually recovered by 12 weeks.

Abnormal habits, such as nocturnal bruxism and daytime clenching, can also be treated with occlusal devices. Okkerse et al. recorded the nocturnal muscular activity of masseter and temporalis anterior muscles, prior and after treatment with a bite-plane. The results showed a significant decrease in nocturnal parafunctional muscle activity with the bite-plane in situ. Similar results, showing the efficacy of occlusal devices in the treatment of occlusal parafunction, were demonstrated with electromyographic examination by Holmgren et al.. The occlusal plate is paramount in relaxing muscle activity and thus potentiates and prolongs BoNTA effect on the masseter.

Rauso et al. in a recent study considered ten patients presenting with widened lower face, treated with percutaneous injection of BoNTA; during the pre-treatment investigations the occlusal parameters were also considered. In this study it was noted that different types of malocclusion were associated with masseter hypetrophy, and also that the 80% of the patients had a positive history of nocturnal bruxism and dental signs of it (tooth abrasion, etc.). The treatment was performed not only with BoNTA injection but also with an occlusal device (a

Michigan plate; Figure 3, 4) and pharmacological therapy to restore and relax masticatory muscles. The results of this study showed, comparing pre- and post- electromiographyc examination and pictures, a long lasting result throughout all the follow up (18 months). This findings was explained in that way: the use of an occlusal device (a Michigan plate), after the BoNTA injections, was able to re-educate the masticatory muscles, avoiding a relapse of the initial biting force at the end of the effect of BoNTA.

Pre-Operative Evaluation of the Patient

First of all, as in every medical treatment, medical history of the patient has to be investigated. Exclusion criteria for the treatment are: pregnancy (or pregnancy attempt); breast-feeding; concurrent administration of quinine, calcium-channel blocker, penicillamine, or amynoglicoside antibiotics; neuromuscular disorders (myasthenia gravis and Eaton–Lambert disease); or peripheral neuropaty (diabetes mellitus and alcoholism).

After that, physical inspection should be performed.

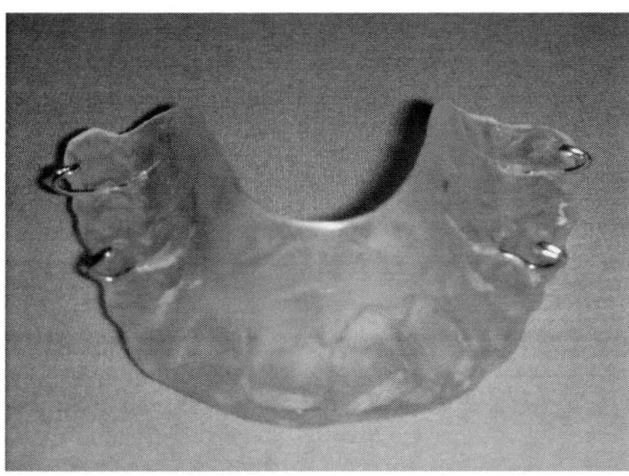

Figure 3. An example of occlusal device: a Michigan plate.

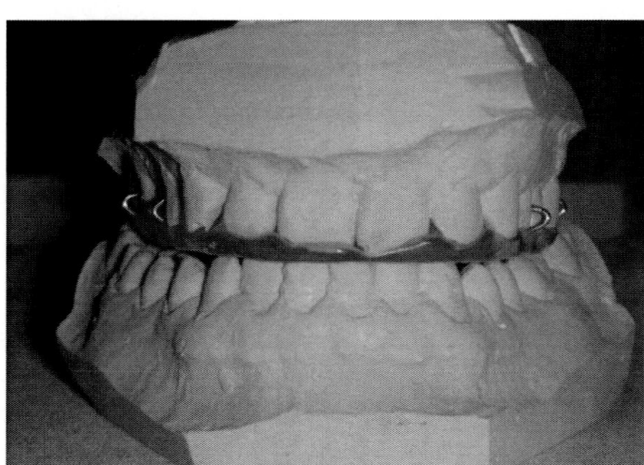

Figure 4. The occlusal device applied on a dental model.

The size of the mandible, the volume of muscle that surrounds the mandible, and the subcutaneous fat tissues determine the width of the lower third of the face. In a case of overdeveloped mandible, it can be corrected by bone resection, and excessive fat tissues can be removed by liposuction. However, if the main factor is masseter hypertrophy, a treatment for the masseter muscle itself is essentially needed. In order to differentiate the diagnosis of skeletal overgrowth, palpation and X-radiography, including panoramic view and the frontal cephalograph, are recommended (Figure 5, 6, 7, 8). Palpation gives inferences regarding the thickness of the muscle and the degree of bone development. However, its not suitable for an objective diagnosis, since the result depends on the physician's subjective experiences. On panoramic view, the physician should primarily examine the location and shape of the condyle, the height of mandibular ramus, and the location of the gonial angle. In particular, the protrusions of developed gonial angles need to be verified by comparing the right and left sides. At this point, we have to consider that the mandibular angle sometimes does not affect the frontal facial aspect, despite the ample appearance on panoramic view. A frontal cephalograph should be taken with more focus on evaluating the developmental status of the gonial angle.

The reason for a wide face without quadrangular bone is not always masseter hypertrophy. The subcutaneous fat tissues also occupy an important portion of the face. In a clinical setting, the pinch test can be used to examine the thickness of the fat layer. In an assessment of the fat layer, the widest portion of the lower face should be identified first. Usually, the bigonial distance is the greatest due to the size of the mandible and the masseter, but sometimes part of the anterior cheek can be selected because of its surrounding thick fat layers.

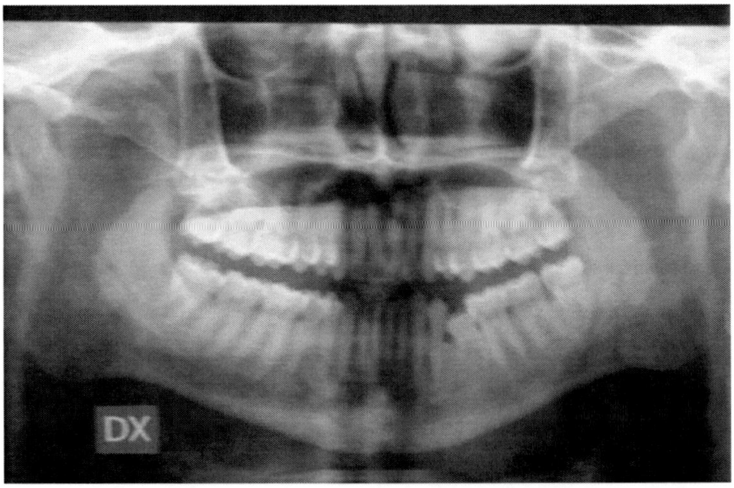

Figure 5. A Panoramic X-Ray showing mandibular angles enlargement.

In the latter case, even if the volume of the masseter muscle is reduced, the width of the anterior cheek is not changed. As a result, it gives a worse impression by causing a depressed look of the mandible. The size of the parotid gland located posterior to the masseter muscle can also affect the width of the lower face. If it is strongly developed, it is dilated more laterally than the masseter muscle. For an objective and simple measurement of the muscle, the subcutaneous fat layer, and the parotid gland, an ultrasonogram can be used. Special attention must be paid when measuring the thickness of soft tissue by ultrasonography: if the physician compresses the target area with the probe, it could be measured as smaller than the actual value.

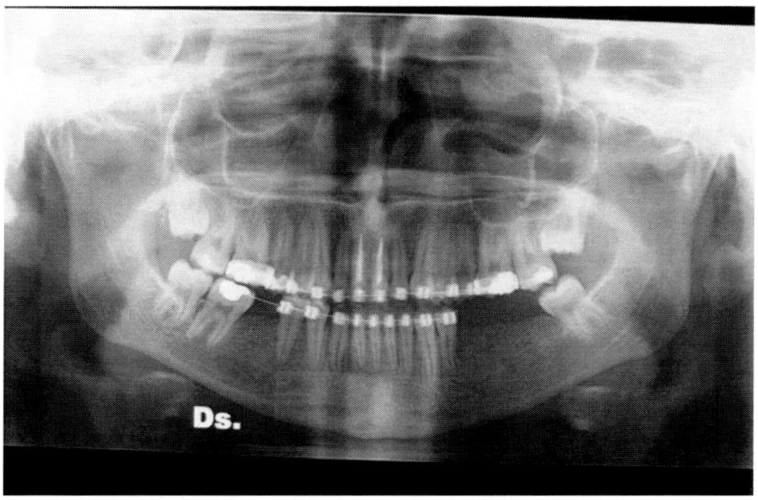

Figure 6. A Panoramic X-Ray with a normal mandibular angles pattern.

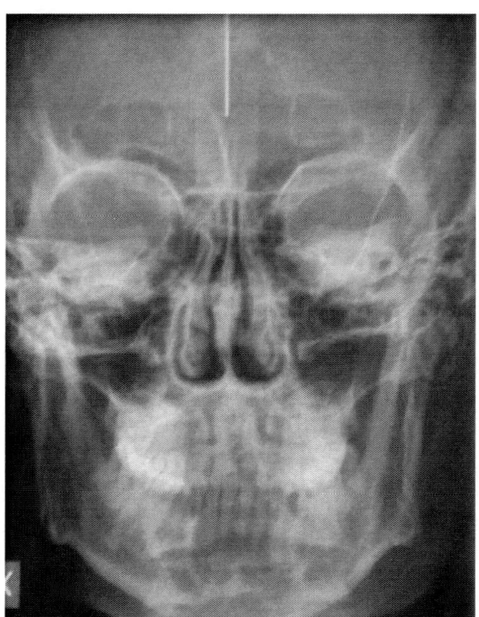

Figure 7. A frontal cephalograph showing mandibular angles enlargement.

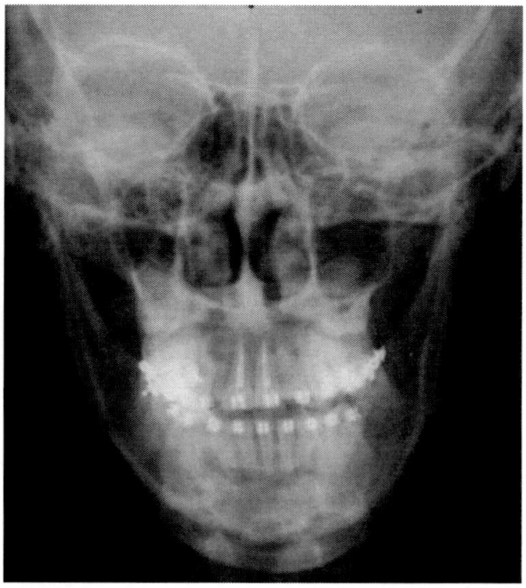

Figure 8. A frontal cephalograph with a normal mandibular angles pattern.

At the least, but not at the last, an-intraoral examination to evaluate dental occlusion, secondary signs of parafunctions (such as tooth abrasion), etc., is mandatory to plan a correct treatment for these patients. It could be very useful to have the dental models of the patient, to have an extensive occlusal evaluation; of course a consultation performed by a dentist is the gold standard.

Usually, patients affected by MMH presents abnormal occlusal habits such as nocturnal bruxism and daytime clenching; those parafunctions can be involved in the genesis of MMH, and are treated with an occlusal device (occlusal splint). Adding an occlusal therapy (using an occlusal device) to BoNTA injections, we can treat extensively the pathology avoiding relapse.

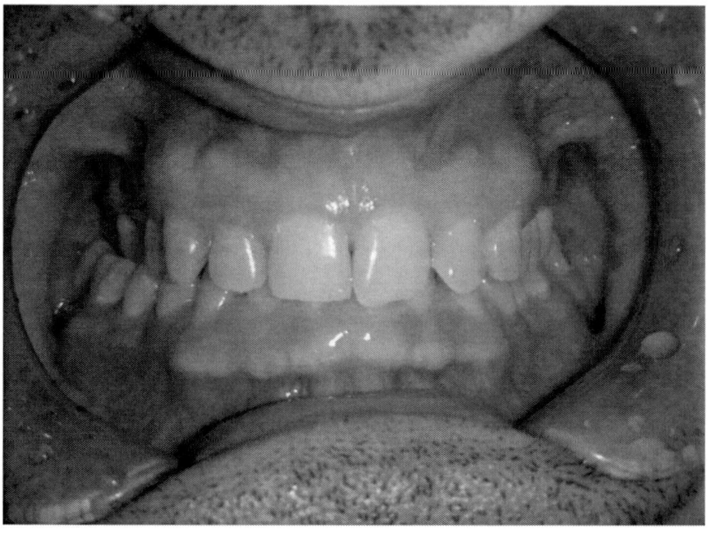

Figure 9. In the figures is shown how dental occlusion changes with a bite plane.

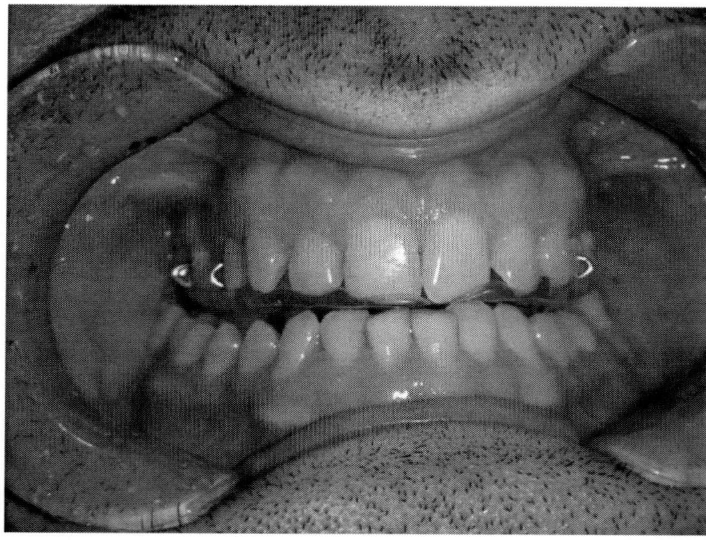

Figure 10. In the figures is shown how dental occlusion changes with a bite plane.

Botulinum Toxin for the Treatment of Masseteric Muscle Hypertrophy

Percutaneous intramuscular injections of BoNTA can be performed to achieve a slim and smooth contour line of the lower face in patients affected by masseteric muscle hypertrophy.

As, underlined before, in the pre-operative evaluation is important to check if the masseteric hypertrophy is associated with occlusal parafunction/disorders, that can be treated with an occlusal device (a splint) (Figure 9; 10). The use of occlusal device should start at the same day of BoNTA injections; at the beginning, for 1 month at least, 16/18 hours per day. Occlusal control must be performed monthly, and at every control the physician should check if the splint contact all the teeth, if not, the splint must be relined. At every control the use of occlusal splint is progressively reduced, 2 hours less every month on average; that way, after 6 months, the patients will use the splint only during night-time. After a couple of months of night-time use (2 to 6 months), its use will be progressively reduced (once every two/three nights), until no more use will be decided by the physician. This, as suggested by Rauso et al., can prolong the effect of the BoNTA, because at when the effect of the neurotoxin finish, the muscles, that have been re-educated by the occlusal device in avoiding excessive contraction, will not have over-contraction anymore, and so a relapse of muscular hypertrophy is avoided.

The drug used for injections is Clostridium Botulinum type A toxin (Botox; Allergan Inc., Dublin, Ireland). Each Botox vial, containing 50 U of BoNTA, has to be diluted with 1.25 mL of normal saline solution with a final concentration of 40 U/mL.

The injection site are marked starting from the inferior margin of the zygomatic arch and the mandibular border; they are the anatomical landmarks to take in mind (young physician can start drawing these areas). The anterior and posterior borders of the masseter muscle are identified by asking the patient to clench his teeth. A grid of 1-cm intervals is drawn on the skin overlying the low and lateral side of the jaw, and a grid of 2-cm up from the zygomatic arch identify were masseteric muscle should be injected (Figure 11, 12). Injections must be performed on the crossed points, with use of a 1.0-mL syringe and a 32-G, 0,7 cm lenght needle; before the injections, it is important to clean the target area using sanitary cotton immersed in alcohol. The toxin has to be administered percutaneously; four injections must be given to each masseteric muscle, (Figure 13) for a total of 16 U of BoNTA in each masseter (Figure 14; 15; 16; 17).

Specific recommendations are not required after the procedure. Patients can claim the reduction of biting force, especially from the fourth to the eighteenth days in the post-op, during mastication; however, this do not compromise the clench ability of the patient.

The result can be observed after 3 weeks on average.

Young physicians should be not afraid about the procedures, because there are not major complications after this treatment.

Minor complications, such as ecchymosis or oedema can be observed, but they resolve spontaneously without any medical treatment in four-five days.

If an asymmetry is present after treatment, or the result is not totally satisfying the patient, the procedure can be repeated; however not before 3 weeks after the first treatment; only at the side of the asymmetry, or both sides to improve the result. The same points of the first treatment have to be injected; half-dosage (2U per point) can be sufficient.

Botulinum toxin type A binds the pre-synaptic cholinergic nerve terminals and blocks the release of acetylcholine; muscles injected with BoNTA are paralyzed and become weak within 2 to 20 days. Although the binding is permanent, muscle contraction usually recovers over 2 to 4 months as new terminal axons sprout, restoring neuromuscular transmission. For that reason the treatment of masseteric muscle hypertrophy with BoNTA is not permanent, however some studies in medical literature affirm that this treatment leads to temporary partial denervation at the area of the neuromuscular synapses and eventually causes muscle atrophy with long lasting results.

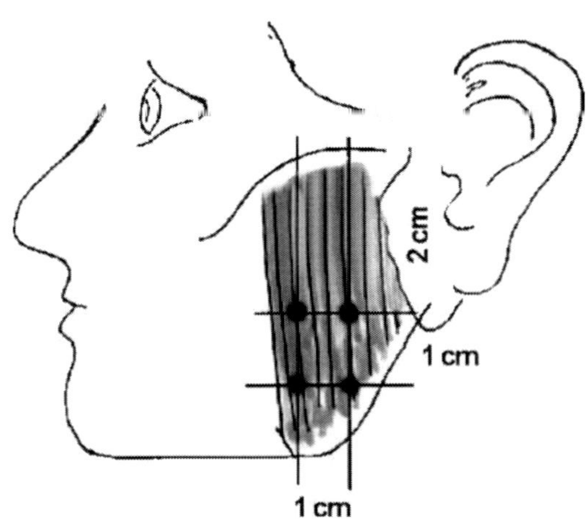

Figure 11. BoNTA injection points for MMH treatment.

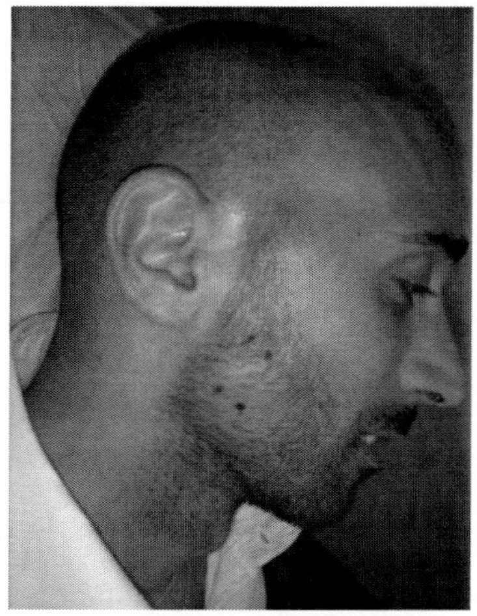

Figure 12. The injection points marked on a patients.

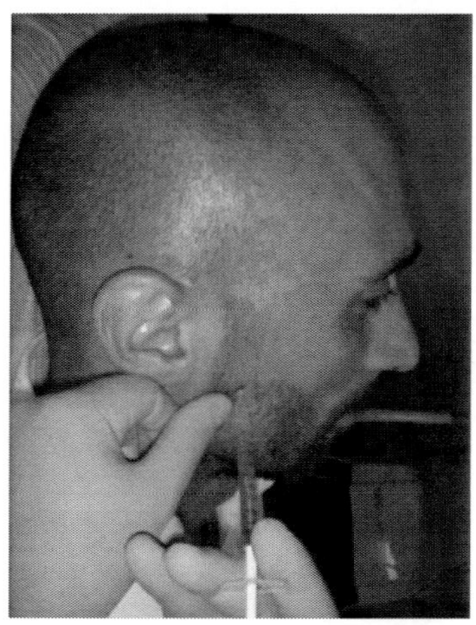

Figure 13. BoNTA injections for MMH.

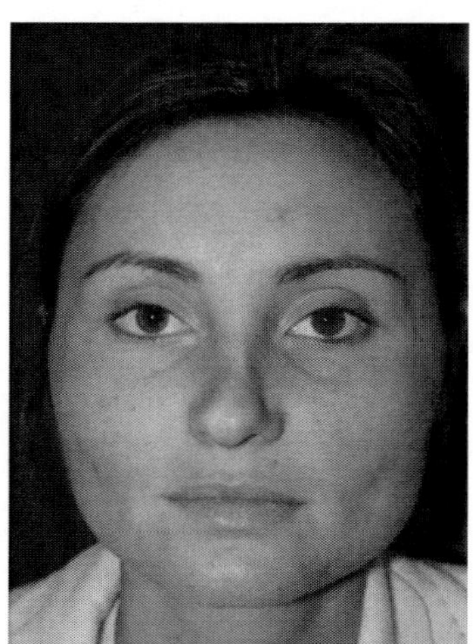

Figure 14. A female patient affected by MMH.

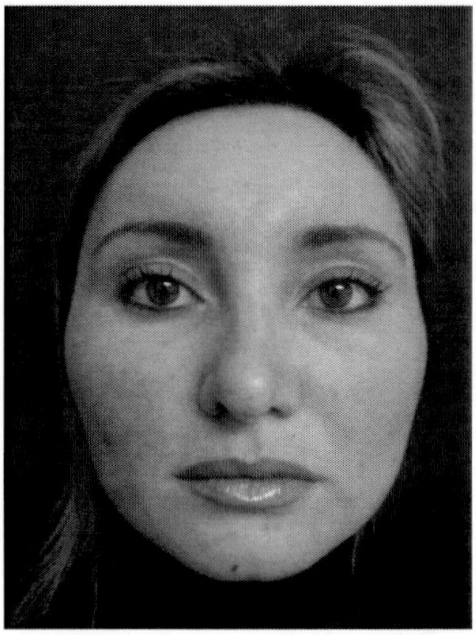

Figure 15. The patient 1 month after the injections.

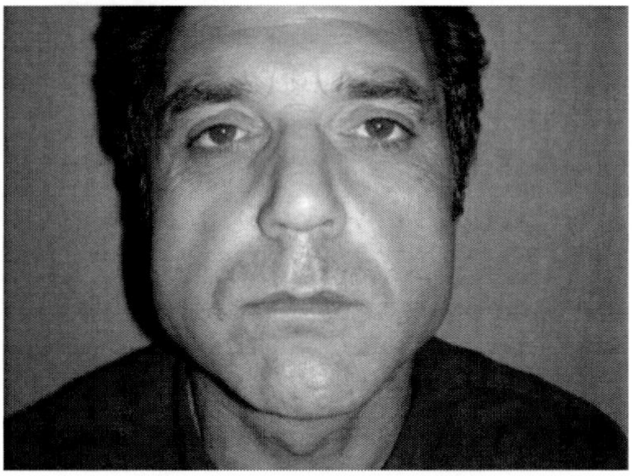

Figure 16. A male patient affected by MMH.

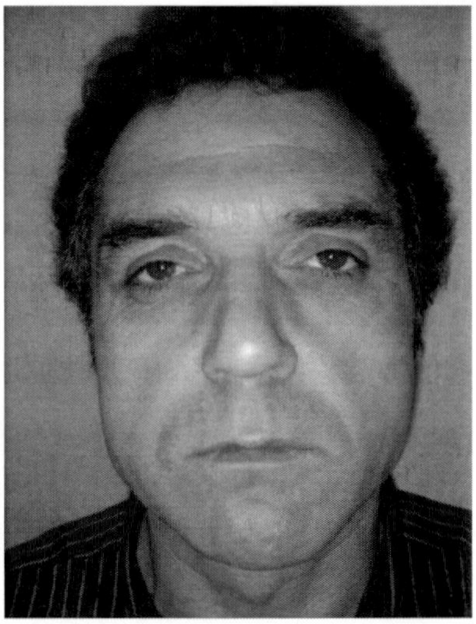

Figure 17. The patient 16 months after injection of BoNTA and occlusal therapy with bite plane.

In clinical practice the effect of BoNTA lasts from 3 up to 7 months; usually the effect is longer in female, probably because masseteric muscle in women is usually smaller and weaker than in men.

Why should We Use Botulinum Toxin Injection in Masseteric Hypertrophy?

It is a therapy that requires simple injection.

Rapid effects take place after treatment, generally appearing in a week.

The injection period is short (about 5 minutes).

There is no inconvenience to daily life after treatment.

There are few side-effects and little pain after treatment.

The cost is inexpensive compared to that of masseter muscle resection or mandible resection.

There is no need to be concerned about permanent side-effects. In 1989, the effectiveness and safety of botulinum toxin A treatment was approved by the Food and Drug Administration (FDA) based on the results of clinical studies done in thousands of patients. Between 1982 and 2000, no cases of side-effects were reported in patients who received botulinum toxin A treatment.

Conclusion

Botulinum toxin type A injection for the contouring of the lower face is simple in technique, has few side effects, and promise a rapid return to daily life; however, it is very important the pre-operative evaluation, including an extensive dental occlusion examination, to plan, according to the dentistry, an extensive therapy to have a long lasting effect, if the masseteric muscle hypertrophy is associated to parafunctional activities or occlusal disorders.

The authors disclose any commercial associations or financial disclosures that might pose or create a conflict of interest with information presented in the manuscript. Such associations include consultancies, stock ownership, or other equity interests, patent licensing arrangements, and payments for conducting or publicizing a study described in the manuscript. Authors disclose any funding received for this work from any of the following organizations: National Institutes of Health (NIH); Wellcome Trust; Howard Hughes Medical Institute (HHMI); and other(s).

References

Ahn KY, Kim ST. The change of maximum bite force after botulinum toxin type A injection for treating masseteric hypertrophy. *Plast. Reconstr. Surg*. 2007;120:1662.

Anonymous. Botulinum toxin. *Lancet* 1992;340:1508-1509.

Baek SM. Aesthetic contouring of the facial skeleton. *Plast. Reconstr. Surg*. 1991;1:673.

Borodic GE, Cozzolino D, Ferrante R, et al. Innervation zone of orbicularis oculi muscle and implications for botulinum A toxin therapy. *Ophthal. Plast. Reconstr. Surg*. 1991;7: 54Y60.

Castro WH, Gomez RS, Oliveira J, et al. Botulinum toxin type A in the management of masseter muscle hypertrophy. *J. Oral Maxillofac. Surg*. 2005;63:20Y24.

Chikhani L, Dichamp J. Bruxism, temporo-mandibular dysfunction and botulinum toxin. *Ann. Readapt. Med. Phys*. 2003; Jul;46(6):333-7.

Choe SW, Cho WI, Lee CK, Seo SJ. Effects of botulinum toxin type A on contouring of the lower face. *Dermatol. Surg*. 2005;31:507.

Converse JM. Masseter Muscle Hypertrophy: Unpublished Case, 1951. In: Reconstructive Plastic Surgery, 2nd ed. Philadelphia: Saunders, 1977.

Dressler D. Botulinum toxin therapy failure: causes, evaluation procedures and management strategies. *Eur. J. Neurol*. 1997;4:S67YS70.

Dressler D. Clinical presentation and management of antibody induced failure in hemifacial spasm. *J. Neurol*. 2004;19:S92YS100.

Finn S, Ryan P, Sleeman D. The medical management of masseteric hypertrophy with botulinum toxin. *J. Ir. Dent. Assoc*. 2000;46:84.

Holmgren K, Sheikholeslam A, Riise C, Kopp S. The effects of an occlusal splint on the electromyographic activities of the temporal and masseter muscle during maximal clenching in patients with a habit of nocturnal bruxism and signs and symptoms of craniomandibular disorders. *J. Oral Rehabil*. 1990;Sept;17(5):447-59.

Hong JU, Choi J, Baek SM. Study on resected area and masticatory muscle's histopatology and radiologic changes after Gato's mandibular osteotomy. *Korea Soc. Plast. Surg*. 1994;21:857.

Jankovic J, Schwartz K. Response and immunoresistance to botulinum toxin injections. *Neurology* 1995;45:1743Y1746.

Jost WH. Other indications of Botulinum toxin therapy. *Eur. J. Neur*. 2006;13: 65Y69.

Kim NH, Chung JH, Park RH, et al. The use of botulinum toxin type A in aesthetic mandibular contouring. *Plast. Reconstr. Surg*. 2005;115: 919Y930.

Klein A, Carruthers A, Fagien S, Lowe N. Comparisons among botulinum toxins: an evidence-based review. *Plast. Reconstr. Surg*. 2008;121:413e.

Lee SK. Antibody-induced failure of botulinum toxin type A therapy in a patient with masseteric hypertrophy. *Dermatol. Surg*. 2007;33: S105YS110.

Mandel L, Tharakan M. Treatment of unilateral masseteric hypertrophy with botulinum toxin: case report. *J. Oral Maxillofac. Surg*. 1999;57:1017.

Monroy PG, da Fonseca MA. The use of botulinum toxin-a in the treatment of severe bruxism in a patient with autism: a case report. *Spec. Care Dentist* 2006; Jan-Feb;26(1):37-9.

Moore AP, Wood GD. The medical management of masseteric hypertrophy with botulinum toxin type A. *Br. J. Oral Maxillofac. Surg*. 1994;32:26.

National Institutes of Health (NIH). Clinical use of botulinumtoxin. National Institutes of Health Consensus Development Conference Statement, November 12Y14, 1990. *Arch. Neurol*. 1991;48:1294.

Newton JP, Cowpe JG, McClure IJ, et al. Masseteric hypertrophy: preliminary report. *Br. J. Oral Maxillofac. Surg.* 1999;37:405.

Okkerse W, Brebels A, De Deyn PP, Nagels G, De Deyn B, Van Bogaert PP, Braem M. Influence of a bite-plane according to Jeanmonod, on bruxism activity during sleep. *J. Oral Rehabil.* 2002;Oct;29(10):980-5.

Pamphlett R. Early terminal and nodal sprouting of motor axons after botulinum toxin. *J. Neurol. Sci.* 1989;92:181Y192.

Park MY, Ahn KY, Jung DS. Botulinum toxin type A treatment for contouring of the lower face. *Dermatol. Surg.* 2003;29:477Y483; discussion 483.

Rauso R, Santagata M, Colella G, Nesi N, Gherardini G, Tartaro G. Can occlusal devices prolong the effect of botulinum toxin type A in the contouring of the lower face? *Eur. J. Plast. Surg.* 2010;33,1:35.

Rijsdijk BA, van ES RJ, Zonneveld FW, et al. De toepassing van botuline A toxine bij cosmetisch storende M-masseterhypertrofie. Ned Tijdschr Geneeskd 1998;142:529.

Rocenvic R. Masseter muscle hypertrophy. Aetiology and therapy. *J. Maxillofac. Surg.* 1986;14:344.

Schwartz KS, Jankovic J. Predicting the response to botulinum toxin injections for the treatment of cervical dystonia. *Neurology* 1990;40:382.

Simpson LL. Peripheral actions of the botulinum toxins. In: Simpson LL, ed. Botulinum neurotoxin and tetanus toxin. San Diego, CA: Academic Press, 1989:153Y178.

Smyth AG. Botulinum toxin treatment of bilateral masseteric hypertrophy. *Br. J. Oral Maxillofac. Surg.* 1994;32:29.

Tan EK, Jankovic J. Treating severe bruxism with botulinum toxin. *J. Am. Dent. Assoc.* 2000; Feb; 131(2):211-6.

Tartaro G, Rauso R, Santagata M, Santillo V, Itro A. Lower facial contouring with botulinum toxin type A. *J. Craniofac. Surg.* 2008;Nov;19(6):1613-7.

To EW, Ahuja AT, Ho WS, et al. A prospective study of the effect of botulinum toxin A on masseteric muscle hypertrophy with ultrasonographic and electromyographic measurement. *Br. J. Plast. Surg.* 2001;54:197.

von Lindern JJ, Niederhagen B, Appel T, et al. Type A botulinum toxin for the treatment of hypertrophy of the masseter and temporal muscle: an alternative treatment. *Plast. Reconstr. Surg.* 2001;107:327.

Whitaker LA. The prominent mandibular angle; Preoperative management operative technique, and results in 42 patients (Discussion). *Plast. Reconstr. Surg.* 1989;83:279.

Yang DB, Park CG. Mandibular contouring surgery for purely aesthetic reasons. *Aesthetic Plast. Surg.* 1991;15:53.

Yu C, Chen P, Chen Y. Botulinum toxin A for lower facial contouring: A prospective study. *Aesth. Plast. Surg.* 2007;31:445-451.

In: Botulinum Toxin
Editors: Fausto E. Marino and Sebastiano De Luca
ISBN: 978-1-62257-185-7

Chapter VI

Botulinum Toxin in Urology: Uses, Mechanisms of Action, Results and Complications

***Cristobal Marchal*[1], *Felipe Saez-Barranquero*[1] *and Maximino Redondo*[2]**

[1]Urology Department, Virgen de la Victoria Hospital, Málaga, Spain
[2]Biochemistry Department, Costa del Sol Hospital, CIBER ESP, Marbella, Spain

Abstract

We analysed the use of botulinum toxin (BoNT) injections as a therapeutic alternative for patients diagnosed with lower urinary tract disorders, reviewing mechanisms of action, results and complications. For this purpose, we evaluated systematic reviews, controlled clinical trials and prospective observational studies in the following urological pathologies: benign prostatic hyperplasia (BPH), overactive bladder (OAB), neurogenic detrusor overactivity (NDO), vesico-sphincter dyssynergia (BND) and interstitial cystitis / painful bladder syndrome (IC / PBS). There is evidence of the effectiveness of treatment with BoNTA in all or some of the symptoms and signs of the OAB, NDO, BND and IC/PBS. However, in BPH there is no clear evidence of the usefulness of this treatment. The main side effect observed was urinary retention.

Aims

To examine the use of botulinum toxin injection (BoNTA) as a treatment alternative for patients diagnosed with lower urinary tract disorders. We review its action mechanisms, results and complications.

Methods

We evaluated systematic reviews, controlled clinical trials and prospective observational studies. The urological diseases studied were benign prostatic hyperplasia (BPH), overactive bladder (OAB), neurogenic detrusor overactivity (NDO), bladder neck dyssynergia (BND) and painful bladder syndrome/interstitial cystitis (PBS/IC).

Action Mechanisms

There are seven types of botulinum toxin (BoNT), of which five are pharmacologically active (A, B, E, F and G). Synthesis of the molecule is based on the binding of a 140-170 kDa chain to a haemagglutinin, creating a complex of 230 to 900kDa. [1]

BoNT has the potential to inhibit the release of multiple neurotransmitters, with its neuronal specificity depending on the type of catalytic domain assigned, and so manipulation of the corresponding light chain is important to the action mechanism. The elimination of nine amino acids from the synaptosomal-associated membrane protein (SNAP-25) by serotype A (BoNTA), produces a more prolonged action [2].

The binding of the SNAP-25 complex to onabotulinum toxin A causes the latter to split. When this occurs the N-ethylmaleinmide-sensitive factor attachment protein receptor complex (SNARE) cannot be formed, and there is no release of the neurotransmitter [3].

BoNTA not only inhibits the presynaptic release of acetylcholine (Ach), but also that of the sensory neuropeptides of substance P (SP). Furthermore, it modifies the expression of subepithelial receptors such as capsaicin receptors (TRPV1) and purinergic receptors ($P2X_3$) which are strongly associated with the human bladder's sensitivity to stretching. It also inhibits the release of ATP, whose action on P2Y receptors also has implications on the mechanisms of bladder pressure, by reducing the sensation of anomalous distension. [4]

The most rapid change observed following treatment with onabotulinum toxin A was the decreased expression of the $P2X_3$ receptors (responsible for the sensation of urgency). Moreover, the blocking of Ach release (which increases with age) has an effect on the parasympathetic nerve endings similar to that provoked by oral anticholinergics. [4]

BoNTA also inhibits the expression of the suburothelial receptors P2Y/SP, causing a decrease in the contractile filaments and gap junctions of human bladder myofibroblasts, which may bring about a reduction in their activation and contractility. [4].

TRPV-1 receptors in the bladder afferent nerve endings can also be activated via neurotrophic growth factors (NGF), which in turn promotes the release of SP via ATP-PY2. Onabotulinum toxin A produces a decrease in the synthesis of NGF, which may lead to peripheral desensitisation. [4]

However, no significant changes were recorded in the histological preparations of the detrusor, which were analysed pre-and post-treatment (axonal and muscle cell degeneration, expanded gap-junctions or intercellular space) [5].

The injection of onabotulinum toxin A into the dog prostate induces apoptotic changes in glands with a predominant stromal component [6] and also decreases the number of inflammatory cells and the expression of COX2 protein, which may produce an antiinflammatory effect [7]. However, with respect to the relaxation of prostate muscle fibres pretreated with onabotulinum toxin A compared to an untreated control group, no statistically significant differences were found [6].

Results

Bladder Neck Dyssynergia (BND)

Most studies of urethral sphincter injections have focused on neurogenic detrusor sphincter dysynergia (nDSD), with level 1 evidence being reported in the two clinical trials published [8,9]. Gallien [9] observed a significant decrease (compared to placebo and a dose of 100 IU of onabotulinum toxin A) in maximum detrusor pressure ($Pdet_{max}$.) and in premictional volume, together with increased bladder capacity (p=0.02). Furthermore, de Sèze [10] reported a significant improvement in post-voiding residue (p=0.01), bladder capacity, maximum urethral pressure and $Pdet_{max}$. (p= 0.02).

The injection of 100 UI of onabotulinum toxin A into 13 patients diagnosed with BND produced a reduction of over 50% of the IPSS score in 87.5% of cases (p=0.02) although these values presented an upward trend from month four. There was also a significant increase in peak urinary flow (Q.Max.), of 3 ml/s at six weeks of treatment (p=0.04) [11]. Moreover, Chen [12] observed a decrease in static and dynamic profiles of urethral closure pressure and in the integrated electromyography (iEMG) (p<0.05).

Other indications are lower urinary tract symptoms without prostatic obstruction [13] and nonneurogenic neurogenic bladder [14].

Benign Prostatic Hyperplasia (BPH)

Of the studies we reviewed concerning treatment with BoNTA for BPH, only two clinical trials [15] and [16] satisfied the requisite quality criteria. The differences in the pre- and post-treatment mean values for the variables Q. Maximum, Prostate Volume, QoL, IPSS and PSA were not statistically significant (p=0.18). Neither was there any statistically significant difference between pre- and post-treatment post-voiding residue (p=0.65). [17]. Marberger [18] found no differences in IPSS score, prostate volume, transition zone volume or Q. Max. between patients treated with 100U-200U-300U of onabotulinum toxin A and those given a placebo. However, with respect to reducing the IPSS score, differences were observed (p = 0.01) between a subgroup previously treated with α-blockers and 200U and the placebo group.

Neurogenic Detrusor Overactivity (NDO)

The main problem encountered in comparisons of the efficacy of BoNTA treatment is the heterogeneity of the condition being treated and the small number of patients included in studies. Thus, it is common to find clinical trials that either do not mention the neurological origin of the lesion [19] or that include spinal cord injury (SCI) and multiple sclerosis (MS) [20] or which group together OB and NDO patients [20]. Of the five controlled clinical trials published [19-23], Grosse [21] is the only researcher who compares abobotulinum and onabotulinum toxin A. In all these cases, there is a significant improvement in continence and in two urodynamic parameters (bladder capacity and compliance), irrespective of the dose used [19].

Del Popolo [24] presents the longest period of monitoring and the highest number of patients (177) in their retrospective cohorts. Different intervals and doses of abobotulinum toxin A were considered, and no significant differences between the treatment groups were observed.

Finally, the mean rate of continence for the different series was 56% (30-87%), with bladder capacity and compliance being the urodynamic parameters presenting the greatest statistically significant improvement ($p<0.001$) [25].

Overactive Bladder (OAB)

With respect to reviews of research in the field of OAB, the main problem is the small number of clinical trials carried out. Thus, even the Cochrane Library [26] includes in its review studies of patients also diagnosed with NDO [20]. Two recent systematic reviews on this topic are identified [27,28], and these report a decreased need for self-catheterisation in 25% of the patients treated [28] and a significant decrease (3.88) in episodes of daily incontinence (CI -6.15, -1.62) [27]. Of the three trials included in this latter review, Brubaker [29] recorded an average duration of 373 days of treatment response. Flynn [30] found no differences between daytime and nighttime urinary frequency, or in Q. Max and $Pdet_{max}$, between the treatment groups. However, Sahai [31] observed a reduction in urinary frequency ($p<0.001$) and urgency ($p=0.008$).

Finally, another trial performed on 242 patients suffering episodes of urge incontinence is currently underway and results will be published this summer [32].

Painful Bladder Syndrome/Interstitial Cystitis (PBS/IC)

The American Urological Association (AUA) recommends, for patients diagnosed with PBS/IC, injection of BoNTA as a fifth line treatment for patients who, following the failure of other treatment options, do not have good control over their symptoms (grade C recommendation) [33].

Giannantoni [34] observed in a systematic review that the overall effect size of onabotulinum toxin A with respect to BoNTA (estimated by the standardized mean difference) is medium (0.2-0.5). However, the effect size on the pain symptom is large (0.8) but minimal with respect to the micturition frequency symptom (<0.2). Kuo [35] found significant differences (using 100 or 200 U of onabotulinum toxin A) against bladder hydrodistention for pain relief (reduction in VAS score of 39% and 55% respectively, p=0.007). Cystometric capacity increased significantly at three months in both treatment groups, compared with bladder hydrodistention (79.5cc and 156.4cc, respectively, p=0.007).

Complications

Overactive Bladder (OAB)

Urinary retention is the complication most frequently reported after injection of BoNTA, and appears to be dose dependent (LoE = 1b) [25]. The overall pooled analysis shows that the BoNTA group is almost nine times more likely to suffer a PVR complication than the placebo group (8.55, 95% CI 3.22-22.71) [27], with male gender (p=0.013) and baseline postvoid residual urine volume (PVR) > 100 cc (p=0.007) being independent predictors of acute urinary tract retention (AUR). Urinary tract infections (UTI) are more frequent in women (p=0.003) and in men with LUTS (p=0.008) [36].

Of the 1025 patients treated with onabotulinum toxin A, only five (0.005%) had muscle weakness, compared to 13 of the 491 patients (0.026%) treated with abobotulinum toxin A [8].

Other side effects include flu-like symptoms [37], dry mouth (10%) and general malaise (5%) [38].

Benign Prostatic Hyperplasia (BPH)

Kuo [15] reported gross haematuria in 145 patients and acute prostatitis in 2%. Following conservative treatment, this was resolved within 3 days. de Kort [39] also reported that two of eleven (18.1%) patients suffered prostatitis.

Neurogenic Detrusor Overactivity (NDO)

Treatment failure attributed to the formation of anti-onabotulinum toxin A [40] or anti-abobotulinum toxin A [41] does not seem significant in current clinical practice. [8].

Painful Bladder Syndrome/Interstitial Cystitis (PBS/IC)

Kuo [35] reported that 46.6% of patients (7 of 15) treated with 100U of onabotulinum toxin A had severe dysuria, with UTI in 20% (3 of 15) and AUR in 13.3% (2 of 15).

Bladder Neck Dyssynergia (BND)

Worsening of preexisting stress incontinence has been reported in 1 of 5 patients (20%) [10] and Gallien observed urinary leakage in 5% of patients [9]. There have also been reports of generalised muscle weakness following urethral injections [42].

Conclusion

There is evidence of the effectiveness of treatment with BoNTA in BND. The two clinical trials examined [8,9] reported a decrease in $Pdet_{max}$ and premictional volume, and increased bladder capacity, following BoNTA injection into the striated sphincter.

In BPH there is no clear evidence of the usefulness of this treatment [17].

In NDO there is a significant improvement in continence (mean 56%) and in two urodynamic parameters (bladder capacity and compliance) [19-23].

The clinical effectiveness in OAB is apparent in the significant reduction (OR 3.88) in daily incontinence episodes [27], but remains more controversial with respect to the decline in daytime and nighttime voiding frequency [30,31].

The most important effect of BoNTA treatment in PBS/IC patients is the decrease in pain symptoms (reduction in VAS score of 39% and 55%, respectively) and the increase in cystometric capacity (of 79.5cc and 156.4cc, respectively) [35]. Nevertheless, the AUA classifies this treatment as fifth line [33].

References

Illie CP, Chancellor MB. Perspective of Botox for treatment of male urinary tract symptoms. *Current Opinion in Urology*. 2009. 19: 20-25.

Pickett A, Perrow K. Towards new uses of botulinum toxin as a novel therapeutic tool. *Toxins*. 2011. 3: 63-81.

Nitti VW. Botulinum toxin for the treatment of idiopathic and neurogenic overactive bladder: State of the art. *Reviews in Urology*. 2006. 8 (4): 198-208.

Apostolidis A, Dasgupta P, Fowler C. Proposed mechanism for the efficacy of injected botulinum toxin in the treatment of human detrusor overactivity. *Eur.Urol*. 2006. 49: 644-650.

Haferkamp A, Schurch B, Krengel U, Grosse J, Kramer G, Schumaker S, Bastian PJ, Bütner R, Müller SC, Stöhrer M. Lack of ultrastructural changes following endoscopic injection of botulinum toxin type A in overactive neurogenic bladder. *Eur.Urol*. 2004. 46: 784-791.

Lin AT, Hang Yan A, Kuo CH. Effects of botulinum toxin A on the contractile function of dog prostate. *Eur. Urol*. 2007.52: 582-9.

Chuang YC, Yoshimura N, Huang CC, Chiang PH, Tyagi P, Chancellor MB. Intraprostatic capsaicin injection as a novel model for nonbacterial prostatitis and effects of botulinum toxin A. *Eur. Urol*. 2007. 51 (4): 1119-1127.

Mangera A, Andersson KE, Apostolidis A, Chapple C, Dagupta P, Giannatoni A, Gravas S, Madersbacher S. Contemporary management of lower urinary tract disease with botulinum toxin A: Systematic review of botox (onabotulinum toxin A) and dysport (abobotulinum toxin A). *Eur.Urol.* 2011. 60:784-795.

Gallien P, Reymann JM, Amarenco G, Nicolas B, de Sèze M, Bellisant E. Placebo controlled randomised, double blind study of the effects of botulinum A toxin on detrusor sphyncter dyssynergia in multiple sclerosis patients. *J. Neurol. Neurosurg. Psychiatry.* 2005. 76: 1670-1676.

de Sèze M, Petit H, Gallien P, de Sèze MP, Joseph PA, Mazaux JM. Botulinum A toxin and detrusor-sphincter dyssynergia, a double-blind lidocaine-controlled study in 13 patients with spinal cord disease. *Eur. Urol.* 42 2002. (1):56-62.

Lim SK, Quek PLC. Intraprostatic and bladder-neck injection of botulinum A toxin in treatment of males with bladder-neck dyssynergia: a pilot study. *Eur.Urol.* 2008. 53: 620-627.

Chen SL, Bih L, Huang YH, Tsai SJ, Lin TB, Kao YL. Effect of single botulinum toxin A injection to the external urethral sphincter for treating detrusor external sphincter dyssynergia in spinal cord injury. *J. Rehabit. Med.* 2008. 40: 744-748.

Chen JL, Chen CY, Kuo HC. Botulinum toxin A injection to the bladder neck and urethra for medically refractory lower urinary tract symptoms in men without prostatic obstruction. *J. Formos. Med. Assoc.* 2009.108 (12): 950-956.

Mokhless I, Gaafar S, Fouda K, Shafik M, Assem A. Botulinum toxin A urethral sphincter injection in children with nonneurogenic neurogenic bladder. *J. Urol.* 2006. 176: 1167-70, discussion 1770.

Kuo HC, Liu HT. Therapeutic effects of add-on botulinum toxin A on patients with large benign prostatic hyperplasia and unsatisfactory response to combined medical therapy. *Scand. J. Urol. Nephrol.* 2009;43:206-211.

Maria G, Brisinda G, Civello IM, Bentivoglio AR, Sganga G, Albanese A. Relief by botulinum toxin of voiding dysfunction due to benign prostatic hyperplasia: results of a randomized, placebo-controlled study. *Urology* 2003; 62:259-264.

Marchal C, Perez JE, Herrera B, Machuca FJ, Redondo M. The use of botulinum toxin in benign prostatic hyperplasia. *Neurourol. Urodyn.* 2012 .31(1):86-92.

Marberger M, Egerdie B, Chartier-Kastler E, Lee KS, Patel A, Bugarin A, Haag-Molkenteller C. A randomised placebo-controlled dose finding study of onabotulinum toxin A in men with BPH. *Eur. Urol.* 2011. Suppl. 10 (2): I-XIV.

Grise P, Ruffion A, Denys P, Egon G, Kastler EC. Efficacy and tolerability of botulinum toxin type A in patients with neurogenic detrusor overactivity and without concomitant anticholinergic therapy: comparison of two doses. *Eur.Urol.* 2010. 58: 759-766.

Schurch B, De Sèze M, Denys P, Cartier-Katler E, Haab F, Everaert K, Plante P, Perrouin-Verbe B, Kumar C, Fraceck S, Brin A. Botulinum toxin type A is a safe and effective treatment for neurogenic urinary incontinence: results of a single treatment, randomized, placebo controlled 6-month study. *J. Urol.* 2005. 174 (1): 196-200.

Grosse J, Kramer G, Jakse G. Comparing two types of botulinum A toxin detrusor injections in patients with severe neurogenic detrusor overactivity: a case control study. *BJU Int.* 2009. 104: 651-656.

Giannantoni A, Di Stasi S, Stephen RL, Bini V, Constantini E, Porena M. Intravesical resiniferatoxin versus botulinum A toxin injections for neurogenic detrusor overactivity: a prospective randomized study. *J. Urol.* 2004.172: 240-243.

Abdel-Meguid T.A. Botulinum toxin injections into neurogenic overactive bladder - to include or exclude the trigone? A prospective randomized, controlled trial. *J. Urol.* 2010. 184: 2423-B.

Del Popolo G, Filocamo MT, Li Marzi V, Macchiarella A, Cecconi F, Lombardi G, Nicita G. Neurogenic detrusor overactivity treated with English botulinum toxin A: 8-year experience of one single centre. *Eur. Urol.* 2008 May;53(5):1013-19.

Apostolidis A, Dasgupta P, Denys P, Elneil S, Fowler C, Giannantoni A, Karsenty G, Schulte-Bauklon H, Schurch B, Wyndaele JJ. Recommendations on the use of botulinum toxin in the treatment of lower urinary tract disorders and pelvic floor dysfunctions: A European consensus report. *Eur. Urol.* 2009. 55: 100-120.

Duthie J, Herbison GP, Wilson D. Botulinum toxin injections for adults with overactive bladder syndrome. *The Cochrane Collaboration.* 2008. 1-16.

Anger J, Weinberg A, Suttorp M, Litwin M, Shekelle P. Outcomes of intravesical botulinum toxin for idiopathic overactive bladder symptoms: a systematic review of the literature. *J. Urol.* 2010. 183(6). 2258-2264.

Cardozo L. Systematic review of overactive bladder therapy in females. *CUAJ* 2011. 5(suppl2): 139-141.

Brubaker L, Richter H, Visco A, Mahajan S, Nygaard I, Braun T, Barber M, Meneefe S, Schaffer J, Weber A, Wei J. Refractory idiopathic urge urinary incontinence and botulinum A injection. *J. Urol.* 2008. 180 (1): 217-222.

Flynn MK, Amundsen C, Perevich M, Liu F, Webster GD. Outcome of a randomized, double-blind, placebo controlled trial of botulinum A toxin for refractory overactive bladder. *J. Urol.* 2009, 181 (6):2608-15.

Sahai A, Khan MS, Dasgupta P. Efficacy of botulinum toxin-A for treating idiopathic detrusor overactivity: results from a single center, randomized, double-blind, placebo controlled trial. *J. Urol.* 2007,177(6):2231-6.

Visco AG, Brubaker L, Richter HE, Nygaard I, Paraiso MF, Menefee SA, Schaffer J, Wei J, Chai T, Janz N, Spino C, Meikle. Anticholinergic versus botulinum toxin A comparison trial for the treatment of bothersome urge urinary incontinence: ABC trial. *Contemp. Clin. Trials.* 2012 33(1):184-196.

Hanno PM, Burks DA, Clemens JQ, Dmochowski RR, Erickson D, Fitzgerald MP, Forrest JB, Gordon B, Gray M, Mayer RD, Newman D, Nyberg L Jr, Payne CK, Wesselmann U, Faraday MM. AUA guideline for the diagnosis and treatment of interstitial cystitis/bladder pain syndrome. *J. Urol.* 2011 185(6):2162-70.

Giannantoni A, Bini V, Dmochowski R, Hanno P, Nickel JC, Proietti S, Wyndaele JJ. Contemporary management of the painful bladder: a systematic review. *Eur. Urol.* 2012, 61(1):29-53.

Kuo HC, Chancellor MB. Comparison of intravesical botulinum toxin type A injections plus hydrodistention with hydrodistention alone for the treatment of refractory interstitial cystitis/painful bladder syndrome. *BJU Int.* 2009, 104(5):657-61.

Kuo HC, Liao CH, Chung SD. Adverse events of intravesical botulinum toxin A injections for idiopathic detrusor overactivity: risk factors and influence on treatment outcome. *Eur. Urol.* 2010 58(6):919-26.

Popat R, Apostolidis A, Kalsi V, Gonzales G, Fowler CJ, Dasgupta P. A comparison between the response of patients with idiopathic detrusor overactivity and neurogenic detrusor overactivity to the first intradetrusor injection of botulinum-A toxin. *J. Urol.* 2005;174(3):984-9.

Ghei M, Maraj BH, Miller R, Nathan S, O'Sullivan C, Fowler CJ, Shah PJ, Malone-Lee J. Effects of botulinum toxin B on refractory detrusor overactivity: a randomized, double-blind, placebo controlled, crossover trial. *J. Urol.* 2005 Nov;174(5):1873-7.

De Kort LM, Rosier P, Kok E et al. Clinical urodynamics and histologic results of intraprostatic injections with botulinum toxin type A for lower urinary tract symptoms due to benign prostatic hyperplasia. *J. Urol.* 2009; 181:703.

Schulte-Baukloh H, Bigalke H, Miller K, Heine G, Pape D, Lehmann J, Knispel HH. Botulinum neurotoxin type A in urology: antibodies as a cause of therapy failure. *Int. J. Urol.* 2008; 15(5):407-15.

Kajbafzadeh AM, Nikfarjam L, Mahboubi AH, Dianat S. Antibody formation following botulinum toxin type A (Dysport) injection in children with intractable bladder hyper-reflexia. *Urology.* 2010; 76(1):233-7.

Pannek J, Göcking K, Bersch U. Long-term effects of repeated intradetrusor botulinum neurotoxin A injections on detrusor function in patients with neurogenic bladder dysfunction. *BJU Int.* 2009; 104(9):1246-50.

In: Botulinum Toxin
Editors: Fausto E. Marino and Sebastiano De Luca
ISBN: 978-1-62257-185-7

Chapter VII

Botulinum Toxin and Facial Palsy

M. L. Navarrete[1], L. Torrent[2], J. Junyent[3], J. F. Fuentes[1], J. R. Moya[1] and A. L. Arellano[4]

[1]Otorhinolaryngology and [2]Rehabilitation Departments,
Vall d´Hebron Hospital, Autònoma University of Barcelona, Spain
[3]Rehabilitation Department, Bellvitge Hospital,
University of Barcelona, Spain
[4]Pharmacology Department, Germans Trias i Pujol Hospital,
Autònoma University of Barcelona, Spain

Abstract

The therapeutic indication of peripheral facial paralysis depends on the degree of nerve injury. A severe facial palsy (electroneuronografic study less than or equal to 10%) develops into a healing sequelae. The sequelae of facial paralysis are the contractures, hemifacial spasm and synkinesis.

Our experience demonstrates that these patients may benefit from rehabilitation treatment (facial exercices and botulinum toxin). We present our work with patients affected by severe peripheral facial paralysis.

They were treated from the beginning of reinnervation by facial exercises according to the Wisconsin School and botulinum toxin. We observe that the subjective and objective efficacy of rehabilitation is high.

Rehabilitation treatment can inform the patient about his chances of recovery, take control and quality of facial expression, greater facial symmetry and thus provide better functionality and quality of life.

Keywords: Facial palsy, rehabilitation treatment, synkinesis, botulinum toxin

Introduction

Charles Bell once wrote: "the human being's facial expression fascinates me; because it serves the most basic and bestial pleasure and participates in the strongest and most gentle emotion of spirit" [1].

Peripheral facial palsy is the paralysis of the muscles innervated by the seventh (VII) cranial nerve.

Its incidence varies between 30-40:100.000 individuals in the general population. Over 100 different causes of Facial Nerve paralisys have been identified. The most frequent causes are: Bell's palsy (up to 50-60 % of all patients), iatrogenic nerve injuries, trauma and both viral and bacterial infections.

In spite of medical and surgical advances, around 30% of patients will have sequelae.

Spontaneous recovery has been observed in around 70% of cases. Incomplete recovery will cause asymmetry, both at rest and with expression, as well as synkinesis with facial mimetism [2].

The static and dynamic changes in the paralyzed face can cause severe functional and psychological disorders, in some cases resulting in permanent disabilities:

An impaired smile can lead to social isolation, and be a factor for depression. Depression symptoms and emotional stress are often associated with facial paralysis and can even augment its disability (40% of patients suffer mild depression symptoms and up to 25% suffer mayor depression) [3].

Difficulty eating and drinking because of incomplete buccal occlusion, especially during fluid intake can also lead to social isolation.

Impaired pronunciation, especially in bilabial consonants like "b", "f", "m", "p" and "s".

Evolution of Facial Paralysis with Partial Recovery

Two different phases can be distinguished:

The initial or hypotonic phase, where there is a lack of function or paralysis of the facial expression muscles. The patient is unable to smile, close the affected eye, frown, or purse the lips.

Partial recovery phase, in which initial movement is evident but facial asymmetry and inability to complete facial movements are present. Afterwards there is post-paralytic synkinesis, sustained muscle contraction and spasms.

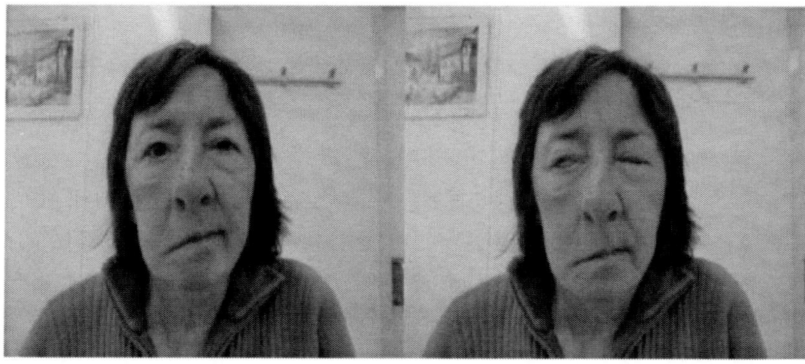

Synkinesis is defined as the involuntary movement of facial muscles provoked by the voluntary movement of other muscles, losing the ability to selectively contract a specific one.

They cause functional limitation during activities like eating, drinking and smiling.

The most frequent synkinesis are: forehead-mouth (the corner of the mouth rises when the eyebrows are raised), eye-mouth (the corner of the mouth rises during eye closure), mouth-eye (the eye closes while smiling, showing teeth or pursing lips), mouth–platysma (neck cords or folds are evident while smiling, showing teeth or pursing lips). Synkinesis is produced by errors in reinnervation [4].

Sustained muscular contraction is defined as muscular tension and rigidity that is frequently observed after neural regeneration.

It is caused by hyperexcitability of the Cranial Facial Nucleus, and happens mostly in dilating muscles.

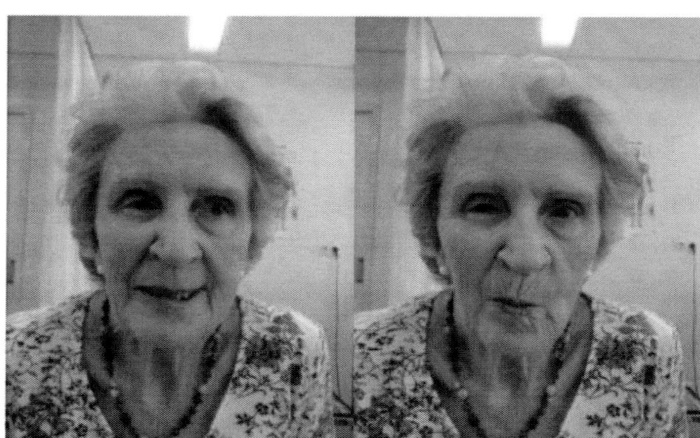

Rehabilitating Treatment

Therapeutic indication of peripheral facial paralysis depends on the degree of nerve injury.

Patients should be referred for rehabilitation treatment:

When a full recovery is not expected. Electroneurography is established as a prognostic factor. It needs to be performed between the 9th-10th day from the onset of paralysis. If < 10% of motor activity is present, recovery will always be incomplete, if it's between 10%-25% recovery will be slow or incomplete and if its 30% or more the prognosis is good.

In case of surgical repair of the Facial Nerve or, after dynamic or static repair surgery.

In case of facial asymmetry, caused by sustained muscular contraction and/or synkinesis.

Prior to starting rehabilitation treatment, the patient is assessed in the outpatient service.

Afterwards, several scoring scales are used on the patient. These should be easy to use, reproducible, and sensitive enough to detect relevant clinical changes:

House-Brackmann Score - A purely motor scale that is divided in 6 grades from normal activity to full paralysis. It gives no information of specific areas or appearance of synkinesis, nor is it sensible to detect small changes during the course of rehabilitative treatment [5].

Facial Grading System or Sunnybrook Test - measures symmetry at rest, symmetry in movement and the appearance of synkinesis. Several statistical analyses have proven the usefulness of this scale in the recovery phase. This is why it is used as a to grade progress during follow-up visits. [6].

Facial Assesment Scale - It is based on a visual appraisal, where voluntary facial movements are scored. It requires a keen and experimented observer, to avoid subjective biases [7].

SAQ - The Synkinesis Assessment Questionnaire involves 10 items, and as its name implies, is used to measure synkinesis [8].

Video recording - useful during the course of treatment, to show and document the patient's progress.

Once a complete medical history and physical examination are done, the need for rehabilitation treatment will be determined based on: the possibility of obtaining symmetry, synkinesis and/or sustained muscular contraction emergence, maintenance and stabilization of motor recovery, the patient's psychological status and his/her phase of recovery.

The aim of rehabilitation treatment for facial paralysis is to improve function and minimize sykinesis and sustained muscular contraction, relative to the degree of injury.

It is not to achieve an "ad integrum" recovery. This is why it's very important for the patient to understand the extent of his injury and his real possibilities of recovery.

Physiotherapeutic treatment is based on achieving: good control and quality of facial mimetism, maximal facial symmetry, acceptable functionality and quality of life. The exercises are progressive and intended to achieve short term objectives, since long term objectives take much longer (between 3-4 years) to complete.

Training cycles of 6 to 8 sessions, each about 60 minutes long, on alternate days are done. A new cycle of training begins when the phase changes.

At the end of each training cycle the patient receives a personalized rehabilitation plan, with a diagram of the facial muscles and an exercise routine that can be completed at home once a day in about an hour. This treatment may be spaced to up to 2-3 times a week, according to the patient's clinical evolution.

Follow up visits should be monthly during the first 3-4 months, every 3-6 months afterward and finally annual.

The physiotherapeutic treatment used is Neuromuscular Re-education. This is a therapeutic approach that uses selective motor training to facilitate symmetry in movement and control synkinesis, varying the exercises according to the phase where the patient is at: hypotonic, reinnervation, Sustained Muscle Contraction/synkinesis [9, 10]. It's based on the plasticity of the CNS to provide new motor patterns.

Treatment based on Neuromuscular Re-education is tailored more to the patient's functional profile than the etiology of the paralysis. It's composed of: movement facilitation techniques, inhibition of anomalous motor patterns and coordination of complex movements.

In spite of the need of personalized treatment, there are common aspects to all patients.

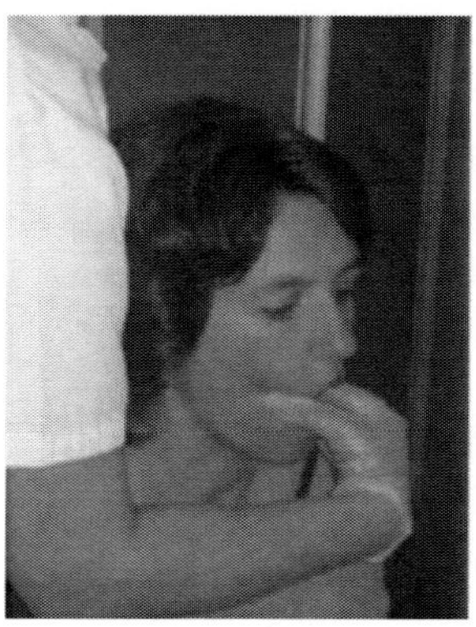

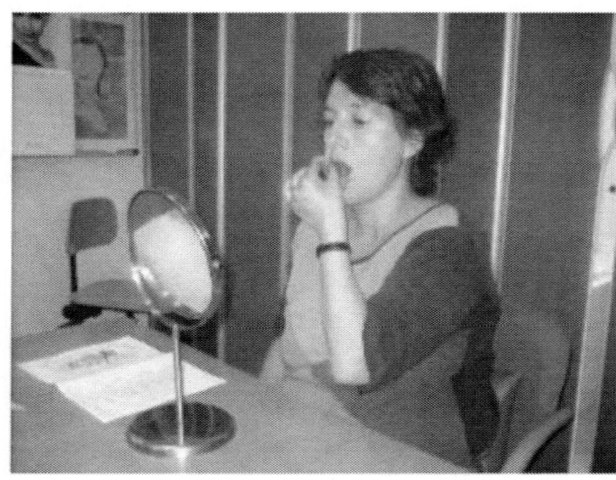

Facial muscle exercises that use feedback to achieve coordination and control of facial mimetism, reduce synkinesis formation and improve strength and range of paretic muscles.

To treat synkinesis, techniques to inhibit aberrant involuntary muscle movement are used. If the inhibition is effective, muscle range, stretching and motor control are gradually increased [11, 12].

Botulinum Toxin in Facial Paralysis Rehabilitation

The use of botulinum toxin in facial paralysis rehabilitation is part of a global treatment, along with neuromuscular reeducation.

The indications for its use are to: symmetrize both sides of the face, control synkinesis and/or sustained muscular contraction [13, 14]. In other cases it's used to control epiphora [15] or hyperhidrosis in the parotid region (Frey syndrome) [16].

Before using botulinum toxin, the patient should first try neuromuscular reeducation. If synkinesis control is not adequate after finishing the training, then infiltration with botulinum toxin should be proposed. Once the procedure is done, the patient is referred once again for reeducation.

After the first infiltration session, the patient is given a follow-up visit in 4 weeks to control the results; if they are satisfactory, a new infiltration is scheduled in 4 months. In advanced phases, the patient is only scheduled for visits every 3-4 months.

Botulinum toxin acts by decreasing the aberrant contractions of facial muscles, blocking acetylcholine at the synapse level. The toxin temporarily paralyzes the target areas of synkinesis during a 4-6 month period.

Combination therapy with both neuromuscular re-education and botulinum toxin results in a "window of opportunity" during which the patient can practice movement patterns without interfering synkinetic movements.

Occasionally after repeated injections, the patient reports a decrease in synkinesis once the toxin's effect has worn off.

Contraindications to botulinum toxin injection are:

Patients with known allergy to the medication.

Generalized muscle function disorders (e.g. myasthenia gravis).

Anticoagulant use (relative contraindication).

Inflammation or infection at the injection site.

Concurrent administration of high doses of amynoglucoside antibiotics (especially in patients with kidney failure)

Pregnancy /Nursing, because of a lack of safety studies.

In our experience, it's also not advisable to inject patients in whom therapeutic follow-up is uncertain or with false "complete cure" expectations.

Before injection, the patient is informed about botulinum toxin: the results we plan to obtain with it, that its use will never create more movement and the possible complications from the treatment.

We dilute 100 IU of botulinum toxin type-A with 2 cc normal saline solution. After swabbing the injection sites with povidone-ionide or alcohol, we infiltrate 2 to 5 IU in each per area (0.05 to 0.1ml) using a 26 gauge needle and a 1ml insulin syringe.

We will infiltrate the muscle groups that form part of the synkinesis and the zones of Sustained Muscle Contraction that alter facial symmetry and cause pain.

We can also infiltrate the healthy hemiface, in those muscles that cause great asymmetry. In this case, only a minimal dose is used so that we don't diminish perception and reeducation may be continued.

Infiltration of more than 30-35 IU en each hemiface should be avoided, because of an increase risk of complications [14]. We do the first infiltrations with minimal doses (2.5 IU) that we'll be varying according to the results obtained.

The synkinesis most frequently treated with botulinum toxin are:

Eye-mouth: Infiltrate the orbicularis oculi at both the internal and external canthus. If there is weak eye closure we should only infiltrate 0.5 IU in 2 points in the lower eyelid and 1 point in the upper eyelid.

Mouth-Eye: Infiltrate the orbicularis oculi at both the external and internal canthus.

Eyebrow-Mouth: Infíltrate the frontal muscle, always bilaterally so we don't hyperactivate the healthy side.

Mouth/Eye-platysma: Infíltrate the middle, anterior, and posterior platysma, along with the digastric/ orbicularis.

SMC: Infiltrate the mentus, procerus – corrugator , lip depressor, levator labii superioris alaeque nasi muscle and depressor septi nasi muscles.

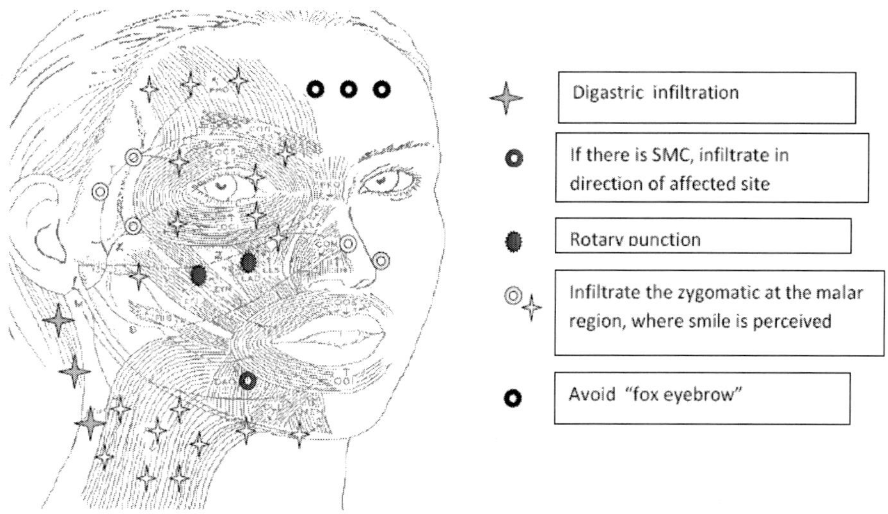

The administration of botulinum toxin at the target muscles provokes a controlled palsy, favoring the reduction of Sustained Muscle Contraction and hampering synkinesis. With all this, we improve facial symmetry, allowing Neuromuscular Re-education without interferences.

Once we infiltrate a new patient, a control visit is scheduled in a week to assess possible complications. A second control at 3-4 weeks is performed to value therapeutic effect, which will serve as reference for the next infiltration. In patients with previous infiltrations, we would only do a 3-4 weeks control to value therapeutic effect.

We will do a new infiltration every 4 months if needed. The infiltration points may differ from previous sessions and less dosage may be required.

Complications

EYE: The most frequent complication is ptosis, seen in up to 7.7% of patients. Less frequent complications include corneal exposure due to difficulty in eye closure, ectropion, lagophthalmos, epiphora and keratitis punctata. Drooping of the eyebrow due to direct infiltration of the frontal muscle has also been described. This can be avoided by infiltrating one finger above the eyebrow.

DROOPING OF THE ORAL COMMISSURE: observed when more than 7 IU of botulinic toxin are infiltrated on the zygomatic major (drooping of the angle of the mouth). The infiltration of the muscles of the cheek or around the lips can cause the patient to bite the inner cheek or to lose the ability to properly pronunciate. This is because the lips are introduced in the mouth.

PAIN: can be decreased by using a finer needle (30 gauge) or by administrating analgesia post-infiltration.

PRURITUS: May occur at the site of injection, described mainly in patients with anticoagulant or antiplatelet treatment. It usually disappears within a week.

FLU- LIKE SYNDROME: observed in between 3.4% to 0.7% of the patients (cervical dystonia) [17].

HEADACHE.

IMMUNITY: the incidence of antibody formation varies in the literature between 3 to 57%. This is mainly because of the different detection techniques used.

The botulinum antitoxin antibodies are frequently present in patients with repeated applications, higher doses or supplementary applications. There appears to be no relation with the patient's age, duration of treatment, number of applications or a total cumulative dose.

The presence of such antibodies hasn't been correlated with the emergence of resistance to treatment, although it is postulated that in some cases it could justify the loss of effectiveness of the toxin.

Special Cases Treated with Botulinum Toxin

Epiphora

A increased in lacrimation [15, 18] due to a taste stimulus is observed in 3.3 to 6.5% of patients with Bell's palsy. It is due to aberrant regeneration of secretory parasympatic fibers of the facial nerve, with an erroneous innervation of the motor-secretory fibers that innerve the submaxillary and sublingual salivary glands through the greater superficial petrosal nerve to the lachrymal gland.

The diagnosis is made with the Schirmer test which is pathologic with less than 15 mm of moisture on the filter paper after five minutes of exposure.

The treatment is done by infiltration of the lacrimal gland of 2 to 2.5 IU of the botilinum toxin either transcutaneous or transconjunctival, with the later one being more effective. A follow-up visit after 10days with a new Schirmer test should be performed.

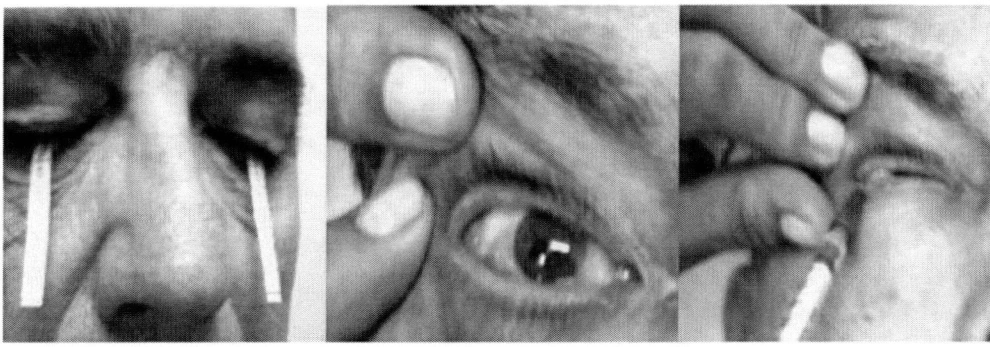

Frey's Syndrome (Auriculo-Temporal Syndrome)

It's defined as an ipsilateral profound preauricular hyperhidrosis during meals (gustatory sweating) and facial flushing [18, 19 and 20]. It is commonly observed within months after parotid gland surgery or injury to the auricotemporal nerve. Diagnosis is made by visual documentation of sweating and can be cuantified with the Minor test or the iodine test after ingestion of food.

Treatment is performed by 6 to 7 injections of 25 to 30 IU of botulinum toxin at the preauricular area. Follow up within 15 days or a month after treatment with a new test.

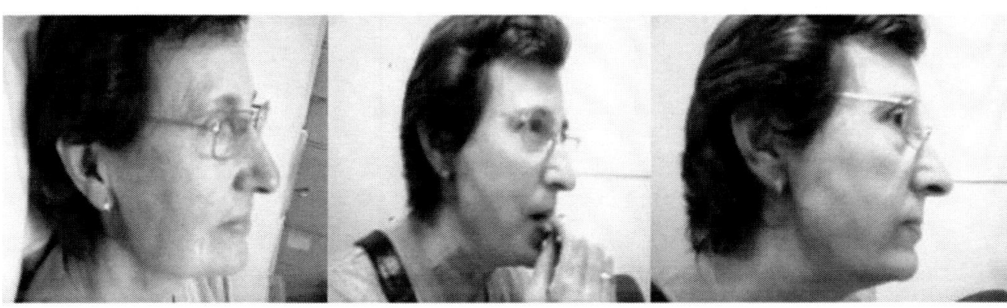

Conclusion

We may conclude that the rehabilitative treatment in facial palsy, that is the association of neuromuscular reeducation with botulinum toxin injections objectively improve synkinesis and SMC and may give the patient: comfort, an improved personal appearance and a more symmetrical and controllable facial function, which will give self-esteem, self-assurance and quality of life. Cientific evidence of Botulinum Toxin shows that it is useful for the treatment of motor and vegetative sequelae of facial paralysis [21].

References

Peripheral facial palsy in the past. Contributions from Avicenna, Nicolaus Friedreich and Charles Bell. Luiz Antonio de Lima Resende1, Silke Weber. *Arq. Neuropsiquiatr.* 2008;66(3-B):765-769.

Recovery of facial movement and facial synkinesis in Bell's palsy patients. Kanaya K, Ushio M, Kondo K, Hagisawa M, Suzukawa K, Yamaguchi T, Tojima H, Suzuki M, Yamasoba T. *Otol. Neurotol.* 2009 Aug;30(5):640-4.

Psychlological distress: Linking impament with disability in facial neuromuscular disorders. VanSwearingen JM, Cohn JF, Turnbull J, Mrzai T, Johnson PC. *Otolaryngol. Head Neck Surg.* 1998, 118:790.

Frequency and location of Synkinesesin Patients with peripheral Facial Nerve Paresis. Carien H.G. Beurskens, Jan Oosterhof, Maria W.G. Nijhuis-van der Sanden. *Otology and Neurotology* 2010. 31:671-675.

Facial Nerve grading System. House H, Brackmann D. *Otoryngol. Head Neck Surg.* 1985, 93: 146-147.

Development of a sensitive clinical facial grading system. Ross BG, Fradet G, Nedzelski JM. *Otoryngol. Head Neck Surgery* 1996, 114:380-386.

Facial assessment scales: defining selective movement. Baliet R, Diels J, Balliet TJ. *Eur. Arch. Otorhinolaryngol.* 1994, S:44-45.

Validation of the synkinesis Assessment Questionnaire. Ritvik p. Metha, Mara WernickRobinson, Tessa A. haddock. *The Larynoscope* 2007, May 117:923-926.

Neuromuscular retraining for facial paralysis. Diels HJ, Combs D. *Otolaryngol. Clin. North Am.* 1997 Oct;30(5):727-43.

Efficacy of functional training of the facial muscles for treatment of incomplete peripheral facial nerve injury. Cai ZG, Shi XJ, Lu XG, Yang ZH, Yu GY. *Chin. J. Dent. Res.* 2010;13(1):37-43.

The effectiveness of neuromuscular facial retraining combined with electromyography in facial paralysis rehabilitation. Cronin GW, Steenerson RL. *Otolaryngol. Head Neck Surg.* 2003 Apr;128(4):534-8.

Physical therapy for Bell s palsy (idiopathic facial paralysis).Teixeira LJ, Soares BG, Vieira VP, Prado GF. *Cochrane Database Syst. Rev.* 2008 Jul 16; (3): CD006283.

Botulinum toxin and facial palsy. Our experience. Navarrete ML, Junyent J., Torrent ML. *Acta Otorrinolaringol. Esp.* 2010 Jul-Aug; 61 (4): 277-81.

Management of synkinesis. Housseman j, Metha RP. *Facial Plast. Surg.,* 2008May; 24(2):242-9.

Treatment of gustatory hyperlacrimation with injection of botulinum toxin into the lacrimal gland. Montoya FJ, Riddell CE, Caesar R, Hague S. *Eye* (2002) 16, 705-709.

Frey's syndrome: treatment with botulinum toxin. DrobikD, Laskawi R. *Acta Otolaryngol.* 1995; 115: 459-461.

Frequency of side effects after botulinum toxin A injections in neurology, rehabilitation and gastroenterology. Slawek j, Madalinski MH, Maciag-Tymecka I, Duzynski W. *Pol. Merkur. Lekarski* 2005 Mar; 18(105):298-302.

The use of botulinum toxin type A in the treatment of Frey and crocodile tears syndromes. Kyrmizakis DE, Pangalos A, Papadakis CE, Logothetis J, Maroudias NJ, Helidonis ES. *J. Oral Maxillofac. Surg.* 2004 Jul;62(7): 840-4.

A review of the surgical and medical treatment of Frey syndrome. Clayman MA, Clayman SM, Seagle MB. *Ann. Plast. Surg.* 2006 Nov; 57(5):581-4.

Management of Frey syndrome. de Bree R, van der Waal I, Leemans CR. *Head Neck.* 2007 Aug;29(8):773-8.

Evaluation and treatment of synkinesis with botulinum toxin following facial nerve palsy. Toffola ED, Furini F, Redaelli C, Prestifilippo E, Bejor M. *Disabil. Rehabil.* 2010;32(17):1414-8.

In: Botulinum Toxin
Editors: Fausto E. Marino and Sebastiano De Luca

ISBN: 978-1-62257-185-7

Chapter VIII

Treatment of Body Deformities Due to Muscle Hypertrophy with Botulinum Toxin A

Luca A. Dessy[1], Marco Mazzocchi[2], Nefer Fallico[1], and Nicolò Scuderi[1]

[1] Department of Plastic and Reconstructive Surgery, Sapienza University of Rome, Rome, Italy

[2] Department of Plastic and Reconstructive Surgery, University of Perugia, Perugia, Italy

Abstract

As cosmetic therapy, botulinum toxin type A (BTX-A) received Food and Drug Administration approval for the treatment of glabellar wrinkles in 2002. Besides this approach, many proven off-label indications of BTX-A are described in the literature.

Patients who practice intense sport activity may develop body deformities ascribable to muscle hypertrophy such as pseudogynecomastia due to the hypertrophy of the pectoralis major muscles, or prominent buttock due to the hypertrophy of gluteus muscles. However, classifications and treatment methods do not take into account such deformities. These forms are particularly evident in body builders with muscles progressively enlarging due to hypertrophy of muscle fibers, but may also be evident in slender individuals practicing non-agonistic sports. In most cases, this aspect is desired and appreciated, but it may be unpleasant for some individuals (i.e. breast with a female appearance or prominent buttock in males). In such cases, consultation with a plastic surgeon may be requested in order to determine the actual presence and nature of the deformity. If it is confirmed that the complained deformity is related to a specific muscle group hypertrophy, a possible solution of the problem is to reduce muscle activity. When a rapid reduction of muscular hypertrophy is requested by the patient, BTX-A injections may represent an option, being an effective non-surgical treatment modality.

Injections of BTX-A are a minimally invasive, very well-tolerated procedure for the treatment of hyperkinetic wrinkles and many other pathologies. The rationale for its use in such deformities consists in the induction of a temporary muscle paralysis by inhibiting the release of acetylcholine at the neuromuscular junction. Usually, one to two sessions of BTX-A injections are sufficient in order to reduce muscle hypertrophy and to obtain the desired result.

In conclusion, BTX-A injections are feasible, cost-effective and repeatable. Furthermore, this treatment achieves evident results with good patient satisfaction, and compliance. The proposed off-label indication should be considered when a rapid resolution of body deformities due to muscle hypertrophy is requested by the patient.

Introduction

Botulinum toxin was first used therapeutically to treat muscular-related pathologies such as strabismus and blepharospasm in the early 80's. It was subsequently observed that its application smoothed facial lines, which prompted researchers to investigate the aesthetic indications for this treatment. By 1992, studies established that botulinum toxin type A (BTX-A) could be used to diminish the appearance of glabellar lines and, in 2002, it received Food and Drug Administration approval for this indication. Besides this approach, many proven off-label indications of BTX-A are described in the literature. Hyperkinetic lines of the rest of the face and of the neck are the most reported off-label cosmetic indications. Furthermore, the use of botulinum toxin A has been employed in improving breast deformity or pain, after pectoralis major (PM) myospasm in sub-pectoral implant breast augmentation or reconstruction. [1-6] The rationale for this treatment is that the botulinum toxin causes a temporary paralysis of the PM muscle, thus allowing a temporary correction of the deformity that several BTX sessions tend to mitigate. Moreover, the use of BTX has been described for the treatment of muscle spasms in several clinical situations such as in irreducible shoulder dislocation, [7] after latissimus dorsi flap and sub-pectoral implant breast reconstruction, [8] and for the treatment of muscle spasm complications of the pectoralis muscle flap in head and neck reconstruction. [9].

Also, the use of BTX-A has been reported to rapidly correct the form of pseudogynecomastia due to the hypertrophy of pectoralis major muscles. [10] Until this latter report, medical classifications and treatment methods did not take into account such deformities. These forms are particularly evident in body builders with muscles progressively enlarging due to hypertrophy of muscle fibers, but may also be evident in slender individuals practicing non-agonistic sports. In most cases, this aspect is desired and appreciated, but it may be unpleasant for some individuals (i.e. breast with a female appearance or prominent buttock in males). In such cases, consultation with a plastic surgeon may be requested in order to determine the actual presence and nature of the deformity.

In this situation, the surgeon should first exclude the presence of a body dysmorphic disorder or dysmorphophobia (from the Greek "dysmorphia"=ugliness). This is a subjective sensation of physical deformity or defect, that the patient feels others notice distinctly, in spite of the fact that his/her appearance is within normal range. When people are affected by body dysmorphism, they often undergo a number of surgical operations, rather than just one, in an attempt to correct this perceived defect. [11,12] This is the first sign of a psychological problem, especially when other individuals who are not involved do not consider such defect as being important or excessive.

This phenomenon is particularly common among individuals who obsessively practice physical exercise, especially body builders, who decide to make use of supplements and pharmaceutical drugs to rapidly increase their muscular mass when beginning their activity or soon afterwards. This type of problem has been called "reverse anorexia", as the obsession is in being perceived as not having sufficient muscular mass. [13] Also, it is rather unlikely that an individual affected by body dysmorphism is fully aware of his/her problem. Therefore, there is a common tendency to ignore the problem itself.

The plastic surgeon often encounters this type of patients. The surgeon's task is to estimate the presence and degree of the body defects claimed by the patient and to understand whether the degree of the defect is proportional to the importance that is given to it. Consequently, the surgeon must decide whether to begin treatment or not, based on the patient's expectations, hereby giving the patient information in terms of realistically obtainable results.

In this chapter, we focus on patients with body deformities as a result of intense sport activity. These deformities involve muscle hypertrophy such as pseudogynecomastia due to the hypertrophy of the pectoralis major muscles, or prominent buttock due to the hypertrophy of gluteus muscles. Lastly, we explore whether these conditions may be treated by BTX-A injections on an outpatient basis.

Pre-Treatment Assessment

Beside a general examination to analyse global health status, the surgeon should investigate the presence of any known or unknown pathologic condition that could be related to the complained defect.During consultation, the surgeon should disclose if the demand for treatment corresponds to the need to solve a real problem, if this is proportional to the defect pointed out or not, and if the chosen treatment is capable of effectively removing the defect, thus satisfying the patient's expectations.An accurate anamnesis and physical examination allow a thorough evaluation of whether the deformities are real and proportional, or absent and disproportional in relation to their actual dimension.Also, it is important to investigate supplements intake (i.e. proteins) and/or pharmaceutical drugs (i.e. estrogens) to reach the desired body appearance. [14]Patients having absent or disproportional defects, despite their complaints, need to be sent for psychological consult, in order to properly evaluate the presence of body dysmorphism disorder diagnostic criteria. [15-19]Specifically, the surgeon should measure the following muscle characteristics:

Grade of muscular hypertrophy. Muscle thickness can be measured by using a caliper and a tailor meter at rest and under maximal voluntary contraction.

Subcutaneous thickness above the involved muscle at rest with a plicometer.

Muscle Hypertrophy Treatment Modalities

If it is confirmed that the complained deformity is related to a specific muscle group hypertrophy, a possible solution of the problem presented to the patient is to reduce muscle activity. This option allows a gradual muscle hypotrophy without the need of any kind of treatment but it has the drawback of needing a reasonable amount of time (several months) to obtain the required result. Often this category of patients that put a particular attention on their body cannot stop to practice intense sport activity and seeks a rapid reduction of the hypertrophy of a specific muscle group. In this situation, BTX-A injections may represent an option, being an effective non-surgical treatment modality.

Patients must be informed about the pros and cons of the BTX- A treatment; in particular, that this is an off-label indication of the use of BTX-A, that it has a temporary effect, and that there is an eventual need to repeat it after several months.

The authors are experienced in using onaboutlinim toxin A (Botox®, Botox® Cosmetic/Vistabel®/Vistabex®, Allergan, Inc., Irvine, CA, USA) to obtain muscle denervation by percutaneous injections, but other types of approved BTX-A may be considered.

Digital photograms of the involved body area should be taken in a standard manner (frontal, lateral and oblique views, at rest and under maximal contraction).

The adopted dilution is 2 mL of normal saline per 50 units of BTX-A (Vistabex®, Allergan, Inc., Irvine, CA, USA), to prepare a solution of 2.5 U/0.1 mL. On each side (breast and buttock), the area to be treated is divided into 20 sectors and each is infiltrated with 0.1 mL of solution using a 30-gauge needle long enough to reach the involved muscle, for a total amount of 50 units for each side. Post-treatment complications and outcomes are recorded.

Clinical follow-ups are set at two weeks after treatment (to evaluate muscle thickness reduction), at three and six months. If further treatment is required, it is repeated after six months.

Clinical Cases

Case 1

A 20-year old patient complaining about a female breast appearance presented to us. One year before, he underwent surgery to remove mammary glands through a Webster approach.

A new consultation was requested because the patient was unsatisfied with the aesthetic result. In particular, he complained about excessive marked contours of PM muscles with persistence of a female breast appearance. This aspect had an important psychological impact on his social and sexual relationships.

The patient also admitted to practicing non-agonistic body-building. Physical examination evidenced the presence of normal glandular and subcutaneous tissues on the breast. A marked hypertrophy of the PM muscles was observed. This aspect was responsible for the complained well-defined bulging and was evident at rest and under contraction of PM muscles (Figure 1).

Ultrasound examination confirmed normal subcutis depth (thickness of 0.9 cm) and the absence of any pathologic glandular tissues. Thus, surgical re-intervention was excluded. Patient was explained that the breast enlargement was mainly a consequence of a muscle hypertrophy due to its hyperactivity. After his refusal to reduce physical exercises on the breast area muscles, the BTX-A off-label indication was explained to the patient, who accepted and fully consented to it. BTX-A (Vistabex®, Allergan, Inc., Irvine, Calif) was diluted as previously mentioned. Each pectoral area was divided into 20 sectors, each one infiltrated with 0.1 ml of solution (Figure 2). During treatment, the patient did not complain about pain, and was instructed to continue his normal activities in the following days. Clinical follow-up after 2 weeks showed an evident reduction of the "muscular" pseudogynecomastia due to the reduction of PM volume, and a more flat breast at rest and under contraction (Figure 3).

The patient was satisfied with the result and was willing to repeat this treatment at the end of the toxin effect.

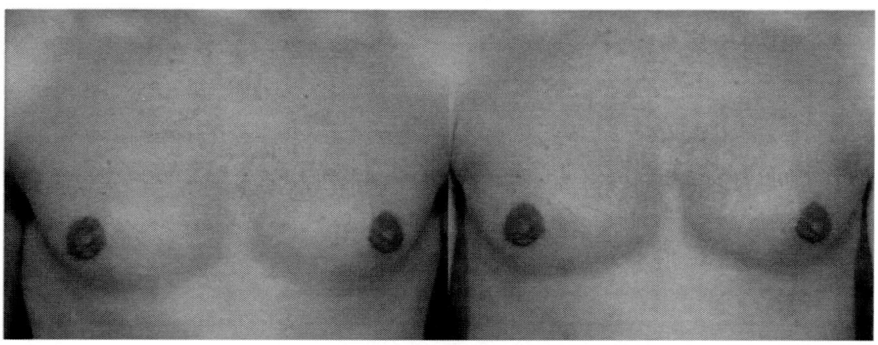

Figure 1. Pre-treatment frontal views of case 1. Left: relaxed position. Right: with contracted pectoralis major muscles.

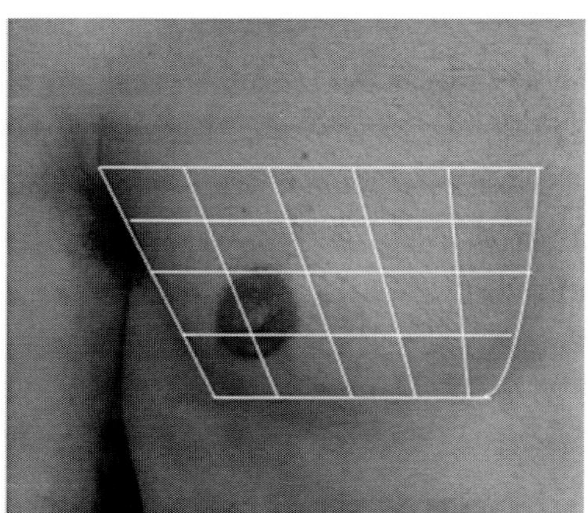

Figure 2. Schematic drawing of the BTX-A injection sites in the pectoral area (20 microinjections of 2.5 U/0.1 mL, for each side).

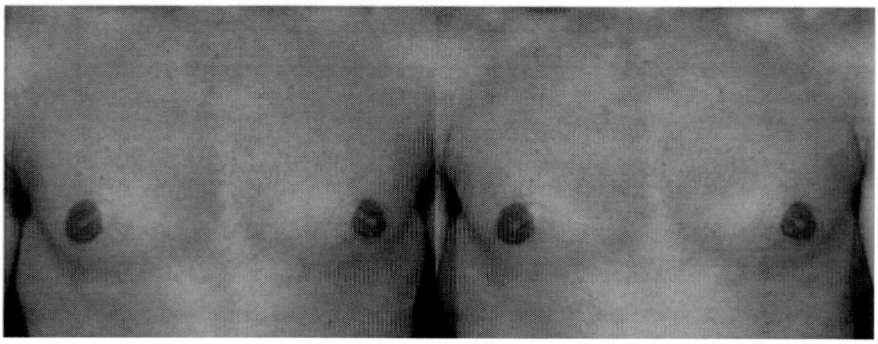

Figure 3. Frontal views 2 weeks after BTX-A treatment of case 1. Left: relaxed position. Right: with contracted pectoralis major muscles.

Case 2

A 28-year old patient complaining about prominent buttocks affecting his psychological status and sexual relationships presented to us. The patient also admitted to practicing non-agonistic body-building. Physical examination evidenced the presence of normal representation of subcutis. A marked hypertrophy of the gluteus muscle groups was observed. This aspect was responsible for the complained defect at rest and under contraction (Figure 4). Any kind of surgery was excluded.

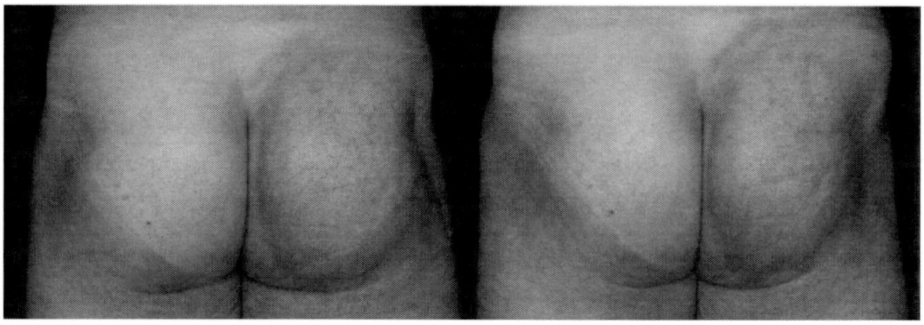

Figure 4. Pre-treatment frontal views of case 2. Left: relaxed position. Right: with contracted gluteus muscles.

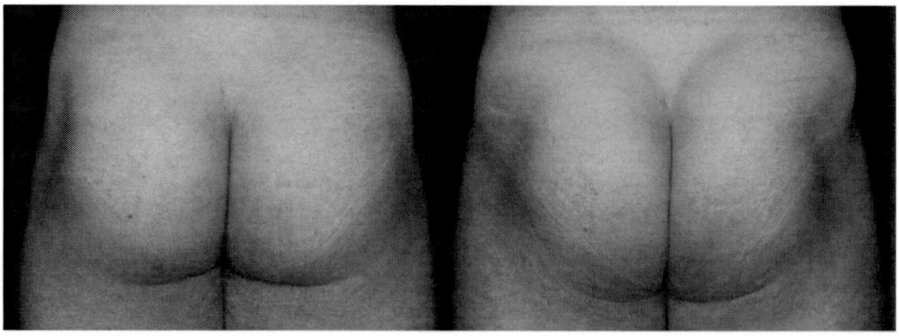

Figure 5. Schematic drawing of the BTX-A injection sites in the gluteus area (20 microinjections of 2.5 U/0.1 mL, for each side).

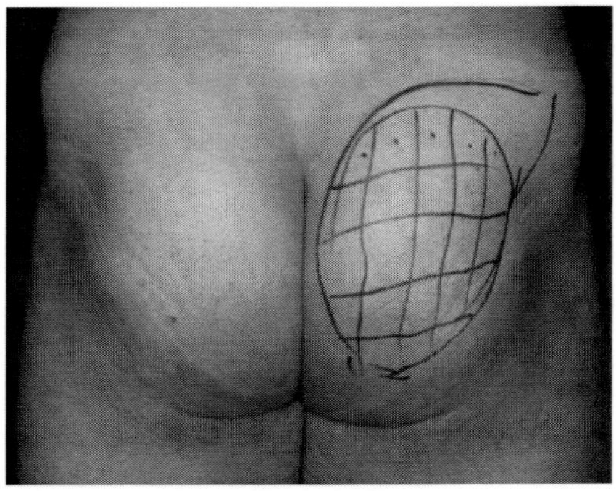

Figure 6. Frontal views 2 weeks after BTX-A treatment of case 2. Left: relaxed position. Right: with contracted gluteus muscles.

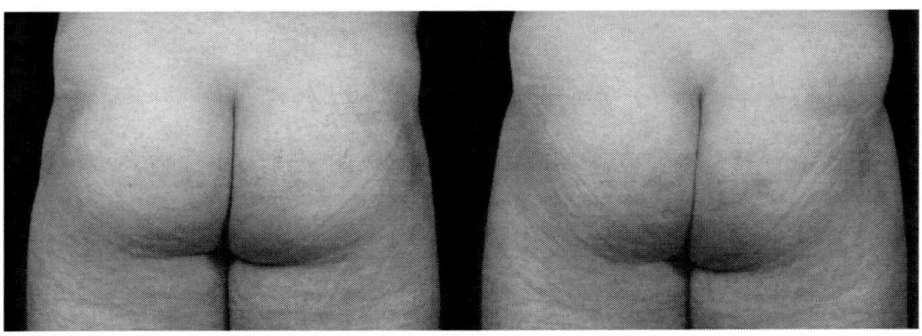

Figure 7. Frontal views 2 weeks after the second BTX-A treatment session of case 2(six months after the first one). Left: relaxed position. Right: with contracted gluteus muscles.

The patient was explained that the defect was mainly a consequence of gluteus muscle hypertrophy. The patient requested a rapid correction of the problem and refused to reduce physical exercises and consented to receiving BTX-A infiltrations. BTX-A (Vistabex®, Allergan, Inc., Irvine, Calif) was diluted as described above. Each gluteus area was divided into 20 sectors, each one infiltrated with 0.1 ml of solution (Figure 5). During treatment, the patient complained about little pain, and was instructed to continue his normal activities in the following days. Clinical follow-up after 2 weeks showed an initial reduction of the defect (Figure 6). The patient requested a further treatment session after six months and he was satisfied with this latest result, (Figure 7) not requesting any further treatment.

Discussion

Deformities deriving from muscle hypertrophy may have an important physical and psychological impact. An accurate diagnostic algorithm is reccomended, with emphasis on a comprehensive history and physical examination. Particular attention should be given to the presence of a body dysmorphism disorder, in order to exclude it.

Current medical classifications do not include deformities due to muscle hypertrophy, i.e. the forms of pseudogynecomastia resulting from the hypertrophy of pectoralis major muscles.

[10] These forms are particularly evident in body-builders because of the progressive hypertrophic enlargement of specific muscle groups. These deformities may also be evident in thin people who practice non-agonistic sports. In the majority of cases, these are desired and appreciated aspects, but sometimes they may be result unpleasant.

The rationale for the use of BTX-A in deformities caused by muscle hypertrophy consists in the induction of a temporary muscle paralysis by inhibiting the release of acetylcholine at the neuromuscular junction.

In order to standardize the procedure, the same amount of BTX-A is used for each treatment session.

The treatment of muscle hypertrophy with BTX-A infiltrations results in a reduction of the muscular tone, allowing a more natural aspect of the area, without the need for a surgical procedure. Although BTX-A causes an irreversible blockade of neurotransmission, its clinical effect is transient. The range of response and efficacy of the BTX-A may depend on the muscle trophism or tone. A study employing different BTX-A concentrations or injected units may be useful to find a relationship between the amount of BTX-A and the severity of the muscle hypertrophy. This may help to better define indications and effectiveness.

Injections of BTX-A are a minimally invasive, low-cost and low-risk procedure for the treatment of muscle hypertrophy. Moreover, They can be easily performed on an outpatient basis. Usually, one to two sessions of BTX-A injections are sufficient in order to reduce muscle tone and to obtain the desired result, avoiding surgical procedures. In addition, it has demonstrated tolerable latency with significant and satisfactory effects already after the first infiltration. The main disadvantage of BTX-A infiltrations is that the result is not permanent and it can be necessary to repeat the treatment in order to obtain the desired result. Nevertheless, it is remarkable that after two sessions the majority of patients demonstrate a considerable improvement and are satisfied with the results.

References

Hoefflin SM. Botox alternatives. *Plast. Reconstr. Surg.,* 1998;101:865.

Wong L. Pectoralis major myospasm resulting from a subpectoral implant. *Plast. Reconstr. Surg.,* 2000;105:1571-1572.

Senior MA, Fourie LR. Botox and the management of pectoral spasm after subpectoral implant insertion. *Plast. Reconstr. Surg.,* 2000;106:224-225.

Richards A, Ritz M, Donahoe S, Southwick G. Botox for contraction of pectoral muscles. *Plast. Reconstr. Surg.,* 2001;108:270-271.

Layeeque R, Hochberg J, Siegel E, et al. Botulinum toxin infiltration for pain control after mastectomy and expander reconstruction. *Ann. Surg.,* 2004;240:608-613.

Cattin T, Govender S. Botulinum toxin for tethering of breast implant. *Plast. Reconstr. Surg.,* 2005;116:686-688.

Sinha A, Higginson DW, Vickers A. Use of botulinum A toxin in irreducible shoulder dislocation caused by spasm of pectoralis major. *J. Shoulder Elbow Surg.,* 1999;8:75-76.

Figus A, Mazzocchi M, Dessy LA, Curinga G, Scuderi N. Treatment of muscular contraction deformities with botulinum toxin type A after latissimus dorsi flap and sub-pectoral implant breast reconstruction. *J. Plast. Reconstr. Aesthet. Surg.*, 2009;62:869-875.

Trignano E, Dessy LA, Fallico N, Rampazzo A, Bassiri B, Mazzocchi M, Scuderi N, Chen HC. Treatment of pectoralis major flap myospasms with botulinum toxin type A in head and neck reconstruction. *J. Plast. Reconstr. Aesthet. Surg.*, 2012;65(2):e23-28.

Dessy LA, Curinga G, Mazzocchi M, Scuderi N. Treatment of muscular pseudogynecomastia with botulinum toxin A. Aesthetic *Plast. Surg.*, 2007;31(1):104-106.

Sarwer DB, Wadden TA, Pertschuk MJ, Whitaker LA. Body image dissatisfaction and body dysmorphic disorder in 100 cosmetic surgery patients. *Plast. Reconstr. Surg.*, 1998;101:1644-1649.

Phillips KA, Dufresne RG, Wilkel CS, Vittorio CC. Rate of Body dysmorphic disorder in dermatology patients. *J. Am. Acad. Dermatol.*, 2000;42:436-441.

Kanayama G, Barry S, Hudson JI, Pope HG Jr. Body image and attitudes toward male roles in anabolic-androgenic steroid users. *Am. J. Psychiatry.*, 2006;163(4):697-703.

Kanayama G, Barry S, Hudson JI, Pope HG Jr. Body image and attitudes toward male roles in anabolic-androgenic steroid users. *Am. J. Psychiatry.*, 2006;163(4):697-703.

Ohjimi H, Shioya N, Ishigooka J. The role of psychiatry in aesthetic surgery. *Aesthetic. Plast. Surg.*, 1988;12:187-190.

Rohrich RJ. The who, what, when, and why of cosmetic surgery: do our patients need a preoperative psychiatric evaluation? *Plast. Reconstr. Surg.*, 2000;106:1605-1607.

Sarwer DB, Crerand CE. Psychological issues in patient outcomes. *Facial. Plast. Surg.*, 2002;18:125-134.

Vargel S, Ulusahin A. Psychopathology and body image in cosmetic surgery patients. *Aesth. Plast. Surg.*, 2001;25:474-478.

Napoleon A. The presentation of personalities in plastic surgery. *Ann. Plastic. Surg.*, 1993;31:193-208.

Webster JP (1946). Mastectomy for gynecomastia through a semicircular intra-areolar incision. *Ann. Surg.*, 124: 557-575.

In: Botulinum Toxin
Editors: Fausto E. Marino and Sebastiano De Luca
ISBN: 978-1-62257-185-7

Chapter IX

Effectiveness of Botulinum Toxin Type A Injection for TMJ Pain and Clicking

Ali Hassani[1],
Mohammad Hosein Kalantar Motamedi[2],
Sarang Saadat[3]
and Mohammad Hamed Mansouri[4]
[1]Oral and Maxillofacial Surgery,
Azad University of Medical Sciences, Tehran, Iran
[2]Oral and Maxillofacial Surgery,
Baqiyatallah University of Medical Sciences, Trauma Research Center, Tehran, Iran
[3]Private Practice Dentisty,Young Researcher Club, Tehran, Iran
[4]Private Practice Dentistry, Tehran, Iran

Abstract

Background and Objective: Temporomandibular joint (TMJ) disorders are among the common causes of head and neck pain; this is also associated with articular crepitus, muscle spasm and movement limitation. Various treatment methods like pharmaceutical therapy, occlusal adjustment, splint therapy and eventually surgery have been used for these patients each having their own advantages and disadvantages. New methods are sought for fast, conservative and complication-free treatment where other treatment modalities fail. At present, use of Botulinum toxin as a new treatment method is under investigation.

Materials and Methods: This prospective interventional study was conducted on 25 patients including 12 females and 13 males suffering from unilateral or bilateral TMJ pain and clicking. These patients had previously received 3 months of anterior repositioning splint therapy with no success. All patients were informed about the treatment with Botulinum toxin and filled out the consent form before initiation of treatment. The study was approved by our ethics committee. Botulinum toxin A was injected into the masseter and temporalis muscles. Volume of injection was 25 units for the temporalis and 50 units for the masseter muscle 5-8 mm deep at three points (anterior, middle and posterior in the temporalis muscle and posterior deep, anterior superficial and inferior in the masseter muscle).

Results: Findings demonstrated that pain scores after 2 weeks and 8 weeks of follow up were decreased it but they did not show a significant reduction (P value=0.1). Minimal changes were detected in terms of clicking (P value=0.1) as well.

Conclusion: In general, this study showed that injection of BTX-A into masseter and temporalis muscles was minimally effective for improvement of clicking and pain.

Keywords: Botulinum toxin type A, TMD, Pain, Clicking

Introduction

TMJ disorders are among important common diseases in the head and neck area and are associated with pain, articular crepitus, muscle spasm, and movement limitation. Various treatment methods like pharmaceutical therapy, occlusal adjustment, splint therapy and eventually surgery have been used for these patients each having their own advantages and disadvantages. Of the disadvantages, we can mention the side effects of drug therapy, long process of splint therapy and invasiveness of surgery [1, 2]. Therefore, it seems necessary to seek and develop new treatment methods for faster complication-free treatment of patients where other treatment modalities fail [1, 3]. At present, use of Botulinum toxin as a treatment method is being considered [3, 4, 5]. BTX-A is a protein complex produced by Clostridium botulinum [6]. The toxin is a 2-chain polypeptide with a heavy and a light chain. The light chain is responsible for the toxicity and the heavy chain is responsible for the binding activity of toxin in cholinergic presynaptic nerve terminals [7]. This toxin inhibits the acetylcholine release at a neuromuscular junction [8]. There is controversy regarding the efficacy of this method. Freund et al. reported satisfactory results after injection of Botulinum toxin into masseter and temporalis muscles [9] while Nixdorf et al, were not satisfied with their results [10]. Considering present controversies, this pilot study aimed at evaluating the effect of Botulinum toxin injection on TMJ clicking and pain.

Materials and Methods

Considering the proven importance of splint therapy, patients underwent 3 months of anterior repositioning splint therapy before initiation of treatment. They used the splint for a minimum of 10 hours at night. If no response to treatment was observed and the pain did not alleviate, BTX-A was injected for them [11]. All patients were informed about treatment with Botulinum toxin and filled out the consent form before initiation of treatment. The study was approved by our ethics committee. This prospective interventional study was performed on 25 patients including 12 females and 13 males complaining of unilateral or bilateral TMJ pain and clicking. Sample selection was based on thorough and standard clinical examination according to research diagnostic criteria [12].

The exclusion criteria were rheumatoid arthritis, psychiatric problems, seizure, pregnancy, neoplastic lesions, use of analgesics in the last 2 weeks and use of other concommitant treatment modalities.

Botulinum toxin type A (Dysport, Ipsen Beaufour Pharmaceuticals, UK) was injected into the muscles (BD Micro-Fine Plus, USA syringe).

Volume of injection was 25 units for temporalis and 50 units for the masseter. All injections were performed by a maxillofacial surgeon. Twenty units of Botulinum toxin was mixed with 0.1 ml normal saline and injected [9]. Needle penetration depth was 5-8 mm at 3 points anterior, middle and posterior in the temporalis and three points posterior deep, anterior superficial and inferior in the masseter as shown in Figure 1. [13, 14].

We set the first follow up session after maximum drug efficacy at 2 weeks and the next after 8 weeks. Clicking and pain were evaluated in the patients. [10, 13-16]. Pain was measured according to visual analogue scale (VAS) which had 0 to 10 scores. Zero meant analgesia, 1-3 mild pain, 4-7 moderate pain, and 8-10 severe pain.

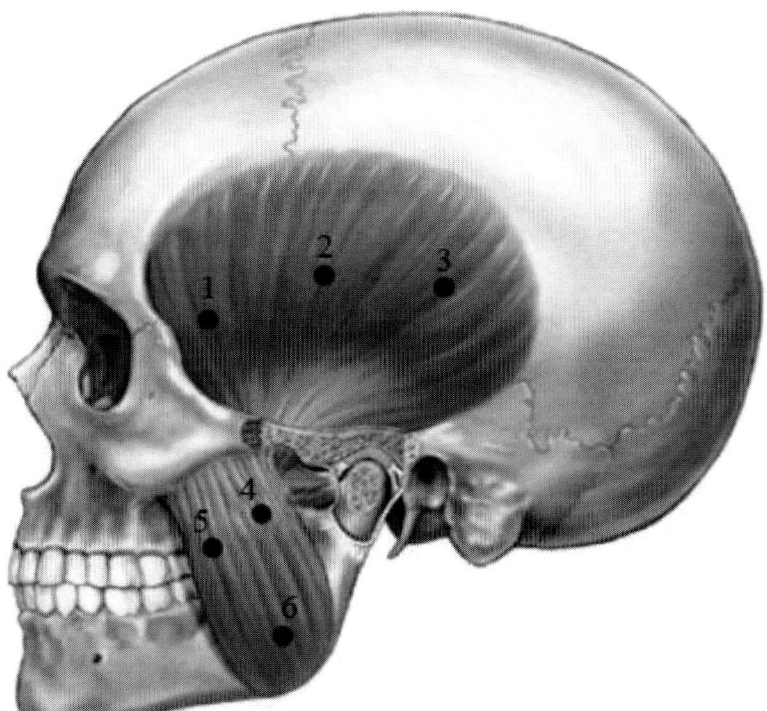

Figure 1. Injection sites. Injection at 3 points anterior (1), middle (2) and posterior (3) in the temporalis and 3 points posterior deep (4), anterior superficial (5) and inferior (6) in the masseter muscle.

Presence or absence of clicking was determined by clinical examination and a stethoscope. Since clicking is a nominal 2-way variable, it was analyzed before and after treatment using the Cochran test. Pain is an ordinal qualitative variable and therefore, it was analyzed using the Repeated Measure ANOVA test.

Results

Based on the results, at the beginning of treatment 4 patients were in the category of mild pain, 14 in the category of moderate and 7 in the category of severe pain. At 2-weeks follow up, 1 had no pain, 6 had mild pain, 13 had moderate and 5 had severe pain. After 8 weeks the results were similar to that of the 2nd week (Table 1). Pain score of the patients in the diagram shows that mean scores of pain were 5.92 before injection, 4.96 at the 2nd week and 4.92 at the 8th week (Table 2).

After statistical analysis P value was calculated to be 0.089 which was not significant. In terms of clicking, before the injection all 25 patients had clicking. In the first follow up after 2 weeks, 21 still had clicking while 4 did not. In the second follow up after 8 weeks 5 did not have clicking and 20 still had it (Table 2). This shows that 84% of subjects in the second week had clicking. This rate was 80% at the 8th week. After statistical analysis P value was calculated to be 0.1 which was not statistically significant.

Table 1. Distribution of patients based on their pain score. Mean pain score of patients before the injection (5.92), at the 2nd week (4.96) and at the 8th week (4.92)

	No pain	Mild pain	Moderate pain	Severe pain
Before injection	0	4 (16%)	14 (56%)	7 (28%)
2 weeks after injection	1 (4%)	6 (24%)	12 (48%)	6 (24%)
8 weeks after injection	1 (4%)	6 (24%)	13 (52%)	5 (20%)

Table 2. Distribution of subjects based on presence or absence of clicking. A total of 84% of patients still had clicking in the 2nd week. This rate reached 80% in the 8th week

	Absence of clicking	Presence of clicking
Before injection	0	25 (100%)
2 weeks after injection	4 (16%)	21 (84%)
8 weeks after injection	5 (20%)	20 (80%)

Discussion

The most common treatment method for TMJ disorders due to masticatory muscles is conservative treatment for resolving the symptoms like use of occlusal splints, physiotherapy, psychological therapy and use of medications [4].

Considering the importance of splint therapy, understudy subjects were selected among those who had not responded to 3 months of splint therapy and we used BTX-A to see if it can be a definitive treatment for this problem.

This study was conducted based on a theory that treatment with Botulinum toxin instead of long-term treatments such as splint therapy and drug therapy is effective for treatment of TMJ problems [4].

BTX is a protein produced by Clostridium botulinum. This protein complex contains a 150 KD neurotoxin connected to a non-toxic component with a non-covalent bond. The latter component is responsible for the stability and protection of the effective component and the final molecular weight of this complex reaches 300-900 KD [6, 17-19]. Composition and molecular weight of the macromolecule depends on the Clostridium botulinum serotype and strain that produces the toxin. BTX types A and B are produced in the laboratory and are fixed, crystallized and dried using products derived from human albumin and are used after dilution with normal saline [8, 19].

The toxin is a two-chain polypeptide with a heavy chain joined by a disulfide bond to a light chain. The heavy chain is responsible for the binding activity of the toxin in cholinergic presynaptic nerve terminals. On the other hand, the light chain is a zinc endopeptidase and is responsible for the toxic properties [7]. This toxin inhibits acetylcholine release from the axon endings but has no effect on signal conduction of the nerve, synthesis or storage of acetylcholine [8, 19].

Intramuscular injection of the proper dosage of BTX in the right location results in partial chemical denervation and decreased muscular contraction without complete paralysis. In glandular tissues, it inhibits the release of secretions [8].

Inhibition of the release of acetylcholine by BTX has several phases. First,irreversible binding of toxin with presynaptic membrane receptors at the neuromuscular junction occurs. These presynaptic receptors are responsible for endocytosis of the neurotoxin at neuromuscular junctions. After endocytosis, the protease present in the nerve terminals breaks the molecule down into 2 polypeptide chains. Variable percentages of the molecules breaking down into chains depends on the serotype. Type A has the highest rate of molecular breakdown (90-95%); while, type B has a lower rate of breakdown of about 70% [17, 18, 19].

The light chain has a high affinity for SNARE proteins (Soluble NFS attachment Protein Receptors). Different targets of this protein also depend on its serotype [18].

Proteolytic breakdown of the SNARE complex by the light chain of BTX prevents the synaptic vesicle from attaching to the internal surface of the cell membrane. Therefore, vesicular attachment and preventing the release of acetylcholine result in development of a mild paralysis in the affected muscle fibers. Its primary effect on the muscle is that it changes the function of the alpha motor neuron responsible for stimulation of muscle fibers. However, BTX can affect gamma motor neurons as well [6, 20].

In this study, injection was done in temporalis and masseter muscles and patients were evaluated for pain and clicking 2 weeks and 8 weeks later. Considering the mechanism of toxicity of BTX it is clear why we chose the mentioned follow up periods. The injected drug will start to function 3-4 days after its injection and reach its maximum efficacy inhibiting all the acetylcholine receptors in 7-10 days. Its effect gradually subsides after 4 weeks and eventually disappearsafter 7 months [4].

The obtained results demonstrated that although pain decreased it was not significant after 2 and 8 weeks (P value=0.1). Although 5 patients were cured of clicking, this was also statistically insignificant (P value=0.1).

However, Symington et al, in 2000 evaluated TMD symptoms of 46 subjects after BTX injection into their masseter and temporalis muscles. A significant decrease ($P<0.05$) was observed in pain, maximum opening and tenderness to touch. Successful outcome of injection in their study might be due to injection at 5 sites in 2 rows in masseter and temporalis muscles [9].

Von Lindern JJ et al, in 2003 assessed pain before and after injection in 3 muscles (masseter, temporalis and pterygoid) in 90 patients. Mean VAS scores significantly improved in 91% of patients ($P<0.01$) [21].

Merete Bakke et al, in their study in 2005 injected BTX into the lateral pterygoid muscles of 2 patients following which, clicking fully resolved and disc position and disc-condyle relationship improved [22].

Nixdorf et al, in 2002 injected Botox into masseter and temporalis and evaluated its effect on pain, maximum opening and tenderness to touch of 15 patients. Their findings were in accord with those of ours. No significant change was observed in severity of pain, tenderness to touch and maximum opening [10].

Disagreements between studies and ours may be due to the multifactorial nature of TMD. Causes of TMD can be very simple or complicated. Our study showed that TMJ pain and clicking although improved by BTX-A injection into masseter and temporalis muscles, this was a small number of cases. This may be due to various etiologies.

The cause of TMD may have not been the targeted muscle which falls into the category of wrong case selection and the resultant no response to treatment. Also, lack of efficacy of BTX-A can be due to the resumed function of the new receptor. It seems that lateral pterygoid muscle has the greatest role for development of TMJ pain and clicking. Since the lateral pterygoid was not among the target muscles in this study it is to be evaluated by injecting BTX-A into this muscle using EMG to assess its effect on improvement of TMD symptoms in future studies.

References

Dao TT, Lavigne GJ.Oral splints: the crutches for temporomandibular disorders and bruxism?. *Crit. Rev. Oral Biol. Med.* 1998;9(3):345-61.

Raphael KG, Marbach JJ. Widespread pain and the effectiveness of oral splints in myofascial face pain. *J. Am. Dent. Assoc.* 2001 Mar;132(3):305-16.

Fu KY, Chen HM, Sun ZP, Zhang ZK, Ma XC. Long-term efficacy of botulinum toxin type A for the treatment of habitual dislocation of the temporomandibular joint.Br. *J. Oral Maxillofac. Surg.* 2010 Jun;48(4):28-4.

Guarda-Nardini L, Manfredini D, Salamone M, Salmaso L, Tonello S, Ferronato G. Efficacy of botulinum toxin in treating myofascial pain in bruxers: a controlled placebo pilot study. *Cranio.* 2008 Apr;26(2):126-35.

Kurtoglu C. Effect of botulinum toxin-A in myofascial pain patients with or without functional disc displacement. *J. Oral Maxillofac. Surg.* 2008 Aug;66(8):1644-51.

Dressler D, Saberi FA, Barbosa ER - Botulinum toxin: mechanisms of action. *Arq. Neuropsiquiatr.* 2005;63:180-185.

Purves D, Augustine GJ, Fitzpatrick D et al.Neuroscience. 3 Ed. Massachusetts, Sinauer 2004;110-116.

Unno Ek, Sakato RK, Issy AM. Estudo comparativo entre toxina botulínica e bupivacaína para infiltração de pontos-gatilho em síndrome dolorosa miofascial crônica. *Rev. Bras. Anestesiol* 2005;55:250-255.

Freund B, Schwartz M, Symington JM. Botulinum toxin: new treatment for temporomandibular disorders. *Br. J. Oral. Maxillofac. Surg*. 2000 Oct;38(5):466-71.

Nixdorf DR. BTX-A for persistent myofascial TMD pain: interpreting a small clinical trial with negative results.*Pain*. 2011 Sep;152(9):1944-5.

Madani AS, Mirmortazavi A. Comparison of three treatment options for painful temporomandibular joint clicking. *J. Oral. Sci*. 2011 Sep;53(3):349-54.

Dworkin SF, LeResche L: Research diagnostic criteria for temporomandibular disorders: Review, criteria, examinations and specifications, critique. *J. Craniomandib. Disord. Facial Oral Pain* 6:301, 1992.

Schwartz M, Freund B. Treatment of temporomandibular disorders with botulinum toxin.*Clin. J. Pain*. 2002 Nov-Dec;18(6 Suppl):S198-203.

Freund B; Schwartz M . Temporal relationship of muscle weakness and pain reduction in subjects treated with botulinum toxin *A. J. Pain* 2003;4:159-165.

Song PC, Schwartz J, Blitzer A. The emerging role of botulinum toxin in the treatment of temporomandibular disorders. *Oral Dis*. 2007 May;13(3):253-60.

Freund B, Schwartz M, Symington JM. The use of botulinum toxin for the treatment of temporomandibular disorders:preliminary findings. *J. Oral Maxillofac. Surg*. 1999 Aug;57(8):916-20.

Wenzel RG . Pharmacology of botulinum neurotoxin serotype A. *Am. J. Heath-Syst. Pharm*. 2004;61(suppl 6):s5-10.

Setler PE - Therapeutic use of botulinum toxins: background and history. *Clin. J. Pain* 2002;18(6 suppl):s119-124.

Colhado OC, Boeing M, Ortega LB. Botulinum toxin in pain treatment.*Rev. Bras. Anestesiol*. 2009 May-Jun;59(3):366-81.

Aoki KR, Pharmacology and immunology of botulinum toxin serotypes. *J. Neurol.* 2001;248(suppl 1):3-10.

Von Lindern JJ, Niederhagen B, Bergé S, Appel T. Type A botulinum toxin in the treatment of chronic facial pain associated with masticatory hyperactivity. *J. Oral Maxillofac. Surg*. 2003 Jul;61(7):774-8.

Bakke M, Møller E, Werdelin LM, Dalager T, Kitai N, Kreiborg S. Treatment of severe temporomandibular joint clicking with botulinum toxin in the lateral pterygoid muscle in two cases of anterior disc displacement.*Oral Surg. Oral Med. Oral Pathol. Oral Radiol. Endod*. 2005 Dec;100(6):693-700.

In: Botulinum Toxin
Editors: Fausto E. Marino and Sebastiano De Luca
ISBN: 978-1-62257-185-7

Chapter X

Botulinum Toxin A for Drug-Refractory Trigeminal Neuralgia

***Behnam Bohluli*[1]**
***and Mohammad Hosein Kalantar Motamedi*[2]**
[1]Azad University, Dept. of OMF Surgery, Dental Branch, Tehran, Iran
[2]Baqiyatallah University of Medical Sciences, Trauma Research Center, Tehran, Iran

Abstract

Botulinum toxin type A (BTX-A) can be used to treat migraine and occipital neuralgia. BTX-A can also be used on trigeminal neuralgia (TN) patients who are refractory to medical treatment. Patients suffering from drug-refractory TN have been treated. We inject 50 U reconstituted BTX-A solution at the trigger zones. Symptoms, including pain duration, provoking factors, affected nerve branch, frequency of TN attacks, and severity of pain just before injections are evaluated 1 week, 1 month, and 6 months after injection. The overall response to treatment compared with values at baseline show men and women aged 28-67 years (mean 48.9 y) suffering from TN from 6 months to 24 years improve regarding frequency and severity of pain attacks; in our patients, pain was completely eradicated and there was no need for further medication. In some patients treated, nonsteroidal antiinflammatory drugs are enough to alleviate pain attacks, and some patients again respond to anticonvulsive drugs after injection. All patients develop higher pain thresholds after injections. A significant difference in frequency of attacks before injection and at 1 week, 1 month, and 6 months after injection is seen. Significant improvement in all patients up to 6 months after injection can be anticipated. Complications include transient paresis of the buccal or mandibular branch of the facial nerve . BTX-A is a minimally invasive method that can play a role in treating TN before resorting to other more invasive therapies, i.e., radiofrequency and surgery.

Introduction

Trigeminal neuralgia (TN) is a unilateral disorder that is characterized by brief pains, abrupt in onset, and limited to the distribution of 1 or 2 divisions innervated by the trigeminal nerve. The prevalence of this pain is about 1 in 25,000 people, and occurs slightly more in women than men. Patients are usually middle-aged or older. [1] Multiple modalities have been used to alleviate the pain, reduce the intensity and frequency of attacks of TN. Treatment with anticonvulsive drugs still seems to be the first choice; surgical procedures, such as microvascular decompression, stereotactic rhizotomy, percutaneous balloon compression, and other methods, are sometimes indicated. [2]

The wide variety of treatment modalities, possibility of severe complications in some techniques, and presence of some refractory cases demonstrates the need for further research. [3] Botulinum toxin type A (BTX-A) has been used to treat pains such as migraine and occipital neuralgia. [4] We use BTX-A on drug-refractory TN patients.

Diagnosis

Patients must be clinically documented for TN according to the Winn criteria (the presence of unilateral paroxysmal stabbing pain limited to the branches of the trigeminal nerve, tender trigger zones, frequent pain free intervals between pain sessions, response to anticonvulsants, and absence of neurologic defect) as well as the criteria defined for TN by the International Headache Society. [5]

Inclusion Criteria

Patients who have had previously undergone pharmacologic treatment protocols (carbamazepine, antiepileptic drugs such as gabapentin, cannabinoids, etc. for several months to a year), but that had become nonresponsive are chosen. Patients with a history of surgery or pathologic conditions (e.g., tumors) are excluded, as are those with a musculoskeletal disorder (a relative contraindication for BTX-A injection).

Treatment

Injection Site

Trigger zones are stated by the patients and confirmed by clinical examination and injection of local anesthesia. Their pain recurs after anesthesia wears off. Patients must be informed of the possibility of transient paresis of the facial nerve at the injection site. Informed consent must be included in the documents of each patient.

Preparation of the BTX-A Injection Solution

After the patients are injected with 1.8 mL lidocaine at the trigger zone, which alleviates the pain while the anesthetic lasts, this confirms the trigger site.

Three milliliters of normal saline solution is mixed in a vial of BTX-A, and a fresh solution is prepared according to manufacturer's guidelines; 50 U reconstituted material is injected at the trigger zones. The effects must be documented at each visit.

Documentation

Documentation must include patient demographics, gender, age, presence of trigger zone, side of involvement, nerve branch involved, total/partial success (after treatment),and complications. Symptoms of pain, including duration, provoking factors, involved nerve branch, frequency of TN attacks (number per day), and severity of pain , just before injections and after 1 week and 1 month post-treatment must also be documented.

Course

Men and women aged 28-67 years suffering from TN from 6 months to 24 years were treated. The mandibular nerve alone may be the origin of pain as well as the maxilla. More than 1 branch of the trigeminal nerve may be involved (both maxilla and mandible, and or the ophthalmic nerve). Patients improve regarding frequency and severity of pain attacks up to 6 months after injection. In many patients, pain may be completely eradicated and there may be no need for further medication. In others, nonsteroidal antiinflammatory drugs may be enough to alleviate pain attacks post-treatment, and patients may again become responsive to anticonvulsive drugs after BTX-A injection. Significant improvement is usually seen in all patients (in the patient global assessment scale) 6 months after injection.

Complications

Complications include transient paresis of the buccal branch or the mandibular branch of the facial nerve. This is usually not significant and resolves in 2 weeks; although paresis may be severe, requiring physiotherapy, however in 3 months usually disappears. We did not have any cases of eyelid ptosis or dysphagia or any systemic side effects.

Discussion

Pain

Idiopathic trigeminal neuralgia occurs in 1 out of 100,000 people and is found more frequently in patients around 50 years of age. It may be typical (i.e., with paroxysmal pain only) or atypical (i.e., with association of a permanent background of pain). Trigeminal neuralgia is an excruciatingly painful neuropathic facial disorder typically presenting as paroxysmal or abrupt pain lasting from several seconds to minutes and rarely up to several hours. The pain is said to feel like stabbing electric shocks occurring spontaneously or after stimulation of a trigger zone. The skin of the face is painful upon TN attacks in the area innervated by V1, V2, or V3 of the trigeminal nerve. An etiology often cannot be found. The pain is severe and may ensue by talking or swallowing. Analgesics are usually ineffective. There may be a trigger point in the oral cavity that sparks the attack. [6] Treatment of TN is possible via both surgery and medication.

Medical Approach

The medical approach is usually used first in an attempt to treat TN noninvasively. This is usually accomplished using anticonvulsants. Carbamazepine is the classic medication of choice for this purpose. Long-term studies, however, have shown a gradual decrease in its efficacy over time. Initial response to this medication is around 80%. After 10 years, however, its effectiveness decreases to 50%.6 Other antiepileptic drugs, such as gabapentin, as well as cannabinoids have also been used.

Surgical Approach

Neurectomy has been reported to be successful in 88.2% of patients. Balloon compression is another method used to treat TN patients, for which initial pain relief prevalence has been reported in 93%. But unilateral facial sensory loss has been reported in 61% of the patients. [6] Use of microvascular decompression (MVD) for TN caused by venous pressure is another effective method of treatment in which the pain recurrence ranges from 31.0% to 75%, within 1-3 years after the initial operation, owing to development of new veins around the nerve root in 87.5% of the cases. Lee et al. [7] did an electronic search of patient records from 1988 to 1998 and found that 393 patients were treated with MVD for TN caused by veins. The pain recurred in 122 patients (31.0%). MVD is a major neurosurgical operation that may have serious complications as well as prolonged convalescence. [6]

BTX-A Injection

The beneficial effects of BTX-A on reducing the frequency and severity of facial pains are documented. Although, the mechanism of this relief is not well understood. Some authors believe that injection of botulinum toxin inhibits secretion of acetylcholine in nerve endings, causing relaxation of muscles and finally relief of pain, whereas others think that the injection stops secretion of some nociceptive neuropeptides in addition to acetylcholine, which may help to prevent pain sensation. [8-10] In our patients who were suffering from refractory TN 50-100 U BTX-A solution was injected at each trigger zone. All experienced considerable relief of pain and some complete cure to date. These findings were in accord with the studies by Türk et al. [11] and Zúñiga et al. [12] In both studies, the pain attacks were considerably alleviated. Türk et al. conducted a relatively similar study, considering in 8 patients who were suffering from TN and injected with 100 U of botulinum solution below and under the zygomatic arch at the involved site. [11] Zúñiga et al. injected BTX into the subcutaneous tissue in 12 TN patients with good results in 10 of them. [12] In Türk et al.'s series no paresis was reported, and complications were dysesthesia in one patient and difficulty in chewing in another. [11] In Zúñiga et al.'s study, paresis was reported in 1 patient that recovered spontaneously. In our study, complete cure was seen in most patients and in the others the severity and frequency of pain were both considerably reduced. In all of these patients, pain threshold was improved and a stronger stimulus was necessary to provoke the pain; pain attacks weakened and duration decreased.

Conclusion

In conclusion, BTX-A has proven to be a safe minimally invasive method that can play a major role in treating TN before other more invasive therapies, (i.e., radiofrequency and surgery). However, long-term assessment is still under way.

References

Katusic S, Williams DB, Beard CM, Bergstralh EJ, Kurland LT. Epidemiology and clinical features of idiopathic trigeminal neuralgia and glossopharyngeal neuralgia: similarities and differences, Rochester, Minnesota, 1945-1984. *Neuroepidemiology* 1991; 10:276-81.

Tatli M, Satici O, Kanpolat Y, Sindou M. Various surgical modalities for trigeminal neuralgia: literature study of respective long-term outcomes. *Acta Neurochir* (Wien) 2008; 150:243-55.

de Siqueira SR, da Nóbrega JC, de Siqueira JT, Teixeira MJ. Frequency of postoperative complications after balloon compression for idiopathic trigeminal neuralgia: prospective study. *Oral Surg. Oral Med. Oral Pathol. Oral Radiol. Endod.* 2006;102:e39-45.

Dodick DW, Mauskop A, Elkind AH, DeGryse R, Brin MF, Silberstein SD. Botulinum toxin type a for the prophylaxis of chronic daily headache: subgroup analysis of patients not receiving other prophylactic medications: a randomized double-blind, placebo-controlled study. *Headache* 2005; 45: 315-24.

Headache Classification Subcommittee of the International Headache Society. The international classification of headache disorders: second edition. *Cephalalgia* 2004; 24(Suppl 1): 9-160.

Motamedi MHK, Rahmat H, Bahrami E, Sadidi A, Navi F, Asadollahi M, Eshkevari PS. Trigeminal neuralgia and radiofrequency. *J. Calif. Dent. Assoc.* 2009;37:109-114.

Lee SH, Levy EI, Scarrow AM, Kassam A, Jannetta PJ. Recurrent trigeminal neuralgia attributable to veins after microvascular decompression. *Neurosurgery* 2000; 46:356-61; discussion 361-2.

Kim DK, Thomas CA, Smith C, Chancellor MB. The case for bladder botulinum toxin application. *Urol. Clin. North Am.* 2006;33:503-10.

Allam N, Brasil-Neto JP, Brown G, Tomaz C. Injections of botulinum toxin type a produce pain alleviation in intractable trigeminal neuralgia. *Clin. J. Pain* 2005; 21: 182-4.

Porta M, Camerlingo M. Headache and botulinum toxin. *J. Headache Pain* 2005; 6:325-7.

Türk U, Ilhan S, Alp R, Sur H. Botulinum toxin and intractable trigeminal neuralgia. *Clin. Neuropharmacol.* 2005; 28:161-2.

Zúñiga C, Díaz S, Piedimonte F, Micheli F. Beneficial effects of botulinum toxin type A in trigeminal neuralgia. *Arq. Neuropsiquiatr.* 2008; 66(3A): 500-3.

Index

A

abnormal habits, ix, 87, 89
acetic acid, 7, 30
acetylcholine, vii, viii, x, 2, 5, 39, 40, 43, 44, 53, 54, 96, 104, 118, 126, 132, 136, 139, 147
achalasia, 2, 29, 88
acid, 3, 56, 72, 83
acidic, 40, 64
acidity, 64
adaptation, 78
adenosine, 7, 30, 44
adenosine triphosphate, 7, 30
adhesions, 74
adjustment, x, 135, 136
adults, 2, 3, 14, 34, 39, 110
adverse effects, 60, 62, 64, 66, 74, 76, 81
adverse event, viii, 22, 24, 54
aesthetic, viii, 54, 55, 64, 65, 66, 69, 71, 88, 100, 101, 126, 128, 133
aesthetics, 65
aetiology, ix, 15, 88, 89, 90
afferent nerve, 7, 8, 30, 104
aggregation, 40
airways, 27
albumin, 58, 139
alcoholism, 91
allergic reaction, 27, 65
allergy, 118
amino acids, 104
amyotrophic lateral sclerosis, 27
analgesic, 44, 45, 72, 74, 81
anatomy, 85
anorexia, 127
ANOVA, 137
antibody, 25, 45, 100, 120
anticholinergic, 32, 109
anticoagulant, 120
antiepileptic drugs, 144, 146
antitoxin, 120
Argentina, 57, 62
artery, 38
arthritis, 26
arthrodesis, 77
articulation, 72
aseptic, 72
assessment, 55, 66, 67, 83, 92, 122, 145, 147
asthenia, 60
asymmetry, 55, 89, 96, 114, 118
asymptomatic, ix, 87
ATP, 6, 8, 30, 31, 44, 104
Austria, 57, 62
autism, 100
axons, 96, 101

B

bacillus, 4, 49
back pain, 11
bacteria, 38
bacterial infection, 114
bacterium, vii, viii, 2, 4, 53, 54
base, 16, 24, 34
bed-wetting, 19
Belgium, 2
beneficial effect, 6, 9, 27, 28, 43, 46, 47, 48, 147
benefits, 12, 39, 42, 44, 62
benign prostatic hyperplasia, ix, 10, 103, 104, 109, 111

bilateral, 47
biological weapons, 40
bladder outlet obstruction, 11
bleeding, ix, 78, 87, 89
blepharitis, 60
blepharoptosis, viii, 54
blepharospasm, 2, 39, 62, 68, 126
blood, 27, 38, 39, 43
blood vessels, 43
body fluid, 38
body image, 133
bonds, 41, 54
bone, 89, 92
Botulinum neurotoxin type A (BoNT-A), vii, 1
botulinum toxin (BTx), vii, 1
botulism, 2, 3, 4, 29, 38, 39, 49
bowel, 23
bradykinin, 43
brain, 11, 24, 42, 43, 46, 47, 48
brain stem, 46
Brazil, 65
breakdown, 139
breast augmentation, 126
breastfeeding, 27
breathing, 3, 26, 27
bruxism, ix, 87, 88, 89, 90, 94, 100, 101, 140
BTX smoothed facial lines, viii, 53
burn, 78

C

Ca^{2+}, 50
calcium, 8, 91
cancer, 83
cannabinoids, 144, 146
carbamazepine, 144
carboxyl, 41
case studies, 17
case study, 17
catalysis, 41
catheter, 28
cation, 7, 8, 31
CDC, 40
cell cycle, 80, 85
cell death, 47
cell surface, 5, 68
central nervous system (CNS), vii, 37, 41, 42, 43, 44, 45, 46, 48, 50, 51, 52, 117
cerebral palsy, vii, 37, 40, 48
cervical dystonia, vii, 37, 39, 40, 48, 62, 64, 76, 84, 101, 120
channel blocker, 91
chemical, 54, 63, 139
Chicago, 23
children, 19, 20, 33, 109, 111
chimera, 47, 48
China, 57, 64, 65
classification, 32, 148
cleavage, 4, 5, 6, 8, 40, 43, 45, 54
clinical diagnosis, 3, 15
clinical examination, 17, 137, 144
clinical presentation, 80
clinical symptoms, 25
clinical syndrome, 4
clinical trials, ix, 21, 39, 55, 103, 104, 105, 106, 108
closure, 11, 105, 115, 118, 119
colonization, 4
commercial, 56, 61, 62, 64, 99
common symptoms, 26
communication, 48
community, viii, 71
compliance, x, 13, 106, 108, 126
complications, vii, ix, 12, 13, 26, 27, 84, 96, 103, 118, 119, 126, 128, 144, 145, 146, 147
composition, 34
compression, 43, 80, 144, 146, 147
computer, 55
computer software, 55
conductance, 31
conduction, 139
conflict of interest, 99
consensus, 21, 22, 24, 32, 34, 100, 110
consent, xi, 135, 136, 144
constipation, 3
containers, 3
contaminated food, 4
contour, viii, 87, 89, 95
contracture, 43
contralateral hemisphere, 43
control group, 105
controlled studies, 16, 19
controlled trials, 13, 14, 18, 20, 22, 24, 28
controversial, 108
controversies, 136
conversion rate, 62
cooking, 3
cortex, 42
cortical bone, 89

cosmetic, vii, viii, x, 1, 2, 10, 28, 37, 39, 40, 53, 54, 55, 56, 61, 62, 63, 65, 66, 68, 69, 75, 125, 126, 133
cost, x, 78, 99, 126, 132
cotton, 96
counsel, 26, 27, 28
counseling, 19
covalent bond, 139
cranial nerve, 114
crepitus, x, 135, 136
crises, 82
crocodile, 123
crystalline, 39, 56
culture, 38, 56
cure, 27, 28, 118, 147
cycles, 116
cyclophosphamide, 30
cystitis, 7, 11, 15, 21
cystoplasty, 12, 13
cystoscopy, 15, 16, 24
cytokines, 74
cytoplasm, 54

D

daily living, 76
daytime clenching, ix, 87, 89, 90, 94
deaths, 12
defects, 127
deformation, 74, 75, 78
dementia, 26
denaturation, 85
dentist, 94
depression, 114
depth, 20, 129, 137
dermatology, 133
desensitization, 31, 74
detection, 8, 33, 120
detection techniques, 120
detrusor overactivity (DO), vii, 1, 12
diabetes, 91
diagnostic criteria, 127, 137, 141
diffusion, 60, 62, 64, 68, 74
diplopia, 62
direct action, 42, 43
disability, 114, 122
disclosure, 74
diseases, 38, 41, 49, 104, 136
dislocation, 126, 132, 140
disorder, 21, 126, 127, 131, 133, 144, 146
displacement, 140, 141
dissatisfaction, 133
distress, 122
distribution, 23, 41, 80, 85, 144
dizziness, 60, 62
dorsal horn, 45
dosage, viii, 21, 54, 62, 66, 96, 119, 139
dosing, viii, 22, 53
double blind study, 109
double-blind trial, 32
drawing, 59, 95, 129, 130
drug action, 29
drug therapy, 136, 139
drugs, xi, 25, 32, 48, 76, 127, 143, 144, 145
dyspareunia, 18
dysphagia, 64, 74, 145
dystonia, vii, 3, 11, 37, 39, 40, 42, 44, 46, 48, 51, 57, 62, 64, 76, 84, 101, 120
dysuria, 107

E

ecchymosis, 96
ectropion, 119
education, 117, 118, 119
electromyography, 75, 77, 105, 122
electron microscopy, 55
EMG, 140
emotion, 114
encephalopathy, 65
endometriosis, 17
enemies, 38
engineering, 47, 48, 52
enlargement, ix, 87, 89, 92, 93, 129, 132
epilepsy, vii, viii, 37, 42, 46, 48, 50
ESD, 19
essential tremor, vii, 37, 40
ethics, xi, 136
etiology, 117, 146
Europe, 46, 57, 61, 62, 64
European Union (EU), 64
evidence, vii, viii, ix, 1, 6, 8, 16, 18, 20, 21, 22, 24, 25, 26, 28, 32, 37, 42, 45, 46, 47, 50, 51, 54, 55, 66, 100, 103, 105, 108, 121
evolution, 69, 117
examinations, 141
excitability, 42
excitation, 45
exclusion, 137
exercise, 117

exocytosis, 7, 31, 40, 41, 49, 50, 51
exposure, 119, 120
extravasation, 47, 52

F

facial asymmetry, 29, 81, 114, 116
facial expression, x, 113, 114
facial muscles, 65, 68, 115, 117, 118, 122
facial nerve, ix, xi, 29, 87, 89, 120, 122, 123, 143, 144, 145
facial pain, 85, 141, 147
facial palsy, vii, ix, 74, 113, 114, 121, 122
faecal incontinence, 31
fat, 92, 93
FDA approval, 9, 10
fever, 3, 60
fiber, 30
fibers, x, 45, 59, 120, 125, 126, 139
fibroblasts, 80, 85
fibrosis, 22, 25, 74, 76
financial, 83, 99
first molar, 90
flank, 19
fluid, 21, 22, 114
food, 1, 3, 38, 121
Food and Drug Administration (FDA), x, 2, 3, 9, 10, 14, 28, 39, 56, 61, 66, 67, 68, 99, 125, 126
food-borne botulism, 1
force, 69, 90, 91, 96, 99
formation, 6, 25, 107, 111, 117, 120
formula, 38
fragments, 48
France, 10, 20, 39, 57, 62, 68, 71
functional changes, 25
funding, 99
fusion, 6, 40, 41, 44, 54, 76, 84

G

ganglion, 5, 8, 31, 45
general practitioner, 33
genitourinary tract, 29
Germany, 1, 10, 38, 39, 57, 61, 62, 68
gingival, 80
gland, 93, 120, 122
glutamate, 44, 45, 46, 74, 80
grades, 55, 116
grading, 122
gray matter, 7
growth, 31, 74, 104
growth factor, 31, 74, 104
guidelines, 145
gynecomastia, 133

H

headache, viii, 3, 11, 40, 44, 48, 51, 54, 60, 62, 63, 148
healing, ix, 113
health, viii, 7, 54, 65, 66, 127
health status, 127
height, 92
hematoma, ix, 87, 89
hemisphere, 43
hemispheric asymmetry, 46
heterogeneity, 105
hippocampus, 43
history, vii, 1, 49, 65, 90, 131, 141, 144
host, 38
House, 116, 122
human, vii, viii, 2, 4, 6, 27, 30, 37, 38, 39, 42, 45, 46, 47, 51, 65, 104, 108, 114, 139
human condition, 47
hydronephrosis, 12
hygiene, 3
hyperactivity, vii, 7, 8, 37, 43, 46, 129, 141
hyperhidrosis, 3, 11, 39, 40, 48, 68, 88, 117, 121
hyperkinetic wrinkles, ix, x, 87, 126
hypertrophy, vii, ix, x, 87, 88, 89, 90, 92, 95, 96, 99, 100, 101, 125, 126, 127, 128, 129, 130, 131, 132

I

iatrogenic, 114
ICS, 12, 15, 16
idiopathic, vii, 1, 7, 8, 12, 14, 30, 31, 32, 34, 45, 108, 110, 111, 122, 147
iEMG, 105
image, 133
immune reaction, 25
immune response, 25
immunogenicity, 62
immunoreactivity, 30, 43
improvements, 12, 13, 14, 15, 16, 18, 19, 23, 24
in vitro, 30, 44
in vivo, 42, 44, 45, 46, 55
incidence, 24, 89, 114, 120
indirect effect, 43
individuals, x, 114, 125, 126, 127

induction, vii, viii, x, 53, 126, 132
infection, 12, 15, 19, 60, 63, 118
inferences, 92
inflammation, 7, 8, 31, 40, 43, 44, 47, 48, 80
inflammatory cells, 104
ingestion, 4, 121
inhibition, vii, viii, 5, 6, 7, 8, 9, 45, 46, 47, 48, 51, 53, 80, 84, 117
initiation, xi, 135, 136
injections, viii, ix, x, xi, 2, 7, 13, 14, 15, 16, 18, 19, 20, 22, 23, 24, 25, 26, 28, 29, 30, 32, 33, 34, 42, 47, 51, 53, 54, 55, 59, 63, 64, 72, 73, 75, 76, 78, 79, 80, 81, 84, 87, 91, 94, 95, 96, 97, 98, 100, 101, 103, 105, 108, 110, 111, 118, 121, 122, 125, 126, 127, 128, 132, 137, 143, 145
injuries, 84, 114
injury, ix, 11, 23, 31, 45, 87, 89, 113, 115, 116, 121, 122
insertion, 89, 132
insulin, 56, 118
intercourse, 18
interference, 66
internalization, 5
interstitial cystitis, ix, 8, 9, 10, 16, 31, 32, 34, 103, 104, 110, 111
intervention, 78, 80, 129
intoxication, 5
intracellular calcium, 31
intramuscular injection, ix, 42, 46, 50, 87, 90, 95
iodine, 121
ipsilateral, 121
Iran, 135, 143
Ireland, 72, 95
irradiation, 82, 83, 84
ischemia, 43, 48, 74
isolation, 39, 114
Italy, 53, 56, 87, 125

J

Japan, 64
joint pain, 83
joints, 43, 72

K

Keloid, 78
kidney failure, 118
Korea, 100

L

laminectomy, 77
latissimus dorsi, 126, 133
lead, viii, 5, 46, 47, 71, 74, 78, 80, 104, 114
leakage, 108
lending, 8
lesions, 26, 66, 137
levator, 18, 33, 74, 119
liberation, 80
light, 4, 5, 6, 40, 54, 81, 104, 136, 139
limb weakness, 27
liposuction, 92
local anesthesia, 72, 78, 144
local anesthetic, 80
localization, 46, 77
lordosis, 76, 77
loss of consciousness, 3
lumen, 40

M

machinery, 40, 41
majority, 9, 21, 23, 26, 28, 132
malaise, 20, 27, 107
malocclusion, 90
man, 39
management, 16, 31, 33, 44, 49, 68, 100, 101, 109, 110, 132
mandible, 72, 89, 92, 99, 145
manipulation, 104
manufacturing, viii, 54
mass, 127
masseter, ix, xi, 72, 87, 88, 89, 90, 92, 95, 99, 100, 101, 136, 137, 139, 140
mastectomy, 132
maxilla, 145
measurement, 55, 66, 72, 92, 101
mechanical ventilation, 3
median, 4
medical, vii, viii, xi, 1, 2, 3, 4, 9, 17, 33, 37, 40, 48, 49, 65, 71, 88, 91, 96, 100, 109, 114, 116, 123, 126, 131, 143, 146
medical care, 33
medical history, 91, 116
medication, xi, 2, 3, 12, 14, 16, 80, 82, 118, 143, 145, 146
medicine, 10, 29
mellitus, 91
metalloproteinase, 4

meter, 127
Mexico, 57, 62
mice, 38, 45, 69, 80
microorganism, 38
microorganisms, 38
migraine headache, 40
migraines, 40, 48, 85
migration, 43, 56
model system, 43
modelling, ix, 87
models, 31, 45, 46, 48, 51, 94
modifications, 16, 74, 81
molecular weight, viii, 54, 139
molecules, 139
morbidity, 17, 19
mortality rate, 4
motor activity, 116
motor control, 117
motor neurons, 43, 139
movement disorders, vii, 37, 40, 46
multiple sclerosis, 10, 13, 106, 109
muscle atrophy, ix, 87, 96
muscle contraction, 96, 114
muscle hypertrophy, vii, ix, x, 87, 88, 89, 95, 96, 99, 100, 101, 125, 126, 127, 129, 131, 132
muscle relaxant, 74
muscle spasms, 11, 126
muscles, vii, viii, x, xi, 1, 3, 10, 17, 18, 25, 33, 37, 40, 43, 53, 65, 74, 76, 89, 90, 91, 95, 96, 114, 115, 117, 118, 119, 120, 125, 126, 127, 128, 129, 130, 131, 132, 136, 137, 138, 139, 140, 147
muscular dystrophy, 90
muscular mass, 127
musculoskeletal, 144
myalgia, 17
myasthenia gravis, 27, 91, 118
myofibroblasts, 104
myosin, 34

N

National Institutes of Health, 99, 100
nausea, 60
negative emotions, 65
nerve, ix, xi, 5, 7, 30, 31, 39, 40, 41, 44, 45, 49, 54, 74, 80, 82, 96, 104, 113, 114, 115, 120, 121, 136, 139, 143, 145, 146, 147
nerve fibers, 45
nerve growth factor, 31
nervous system, 38, 41, 42, 46
neuralgia, xi, 80, 81, 82, 85, 143, 144, 146, 147, 148
neurogenic bladder, 13, 19, 33, 105, 108, 109, 111
neurogenic detrusor overactivity (NDO), ix, 2, 103, 104
neurons, 7, 8, 23, 31, 41, 42, 43, 45, 46, 47, 48, 50, 51, 74
neuropathic pain, 45, 48, 80, 85
neuropathy, 47, 52, 74, 84
neuropeptides, 23, 44, 104, 147
neuroprotection, 47, 52
neuroscience, 48
neurotransmission, 5, 132
neurotransmitter, 5, 8, 44, 45, 50, 104
neurotransmitters, 5, 40, 44, 46, 74, 104
neutral, 40, 76
New Zealand, 39
nitrous oxide, 80
nocturia, 12, 14, 16
nocturnal bruxism, ix, 87, 89, 90, 94, 100
nonsteroidal antiinflammatory drugs, xi, 143, 145
norepinephrine, 44
normal children, 33
North America, 46, 57, 61
nuclei, 45
nucleus, 45
nystagmus, 11

O

obstruction, 10, 105, 109
occlusal parafunction, ix, 88, 90, 95
occlusion, 94, 95, 99, 114
oedema, 60, 62, 96
oesophageal, 2
operations, 126
ophthalmologist, 2
opportunities, 48
oral cavity, 146
organs, 31, 47, 74
osteotomy, 100
outpatient, 116, 127, 132
overactive bladder (OAB), ix, 103, 104
ownership, 99
oxygen, 80

P

pain, vii, viii, x, xi, 1, 6, 7, 8, 10, 11, 15, 16, 17, 18, 19, 21, 23, 25, 26, 30, 33, 34, 37, 40, 42, 43, 44, 45, 46, 47, 48, 50, 51, 52, 60, 63, 71, 72, 73, 74,

75, 76, 78, 80, 81, 82, 83, 84, 85, 99, 106, 108, 110, 118, 126, 129, 131, 132, 135, 136, 137, 138, 139, 140, 141, 143, 144, 145, 146, 147, 148
pain management, vii, viii, 37, 42, 48, 50, 51
pain perception, 7, 81
palpation, 78, 92
parallel, 7, 67
paralysis, vii, viii, ix, x, 3, 4, 6, 26, 29, 40, 41, 47, 49, 53, 54, 113, 114, 115, 116, 117, 121, 122, 126, 132, 139
paresis, xi, 143, 144, 145, 147
parietal cortex, 42
parotid, 84, 92, 117, 121
parotid gland, 84, 92, 121
pasteurization, 3
pathogenesis, 76
pathology, 6, 10, 12, 15, 71, 80, 94
pathophysiological, 30
pathophysiology, 9, 23, 46
pathways, 16, 23, 24, 30, 44
pectoralis major, x, 125, 126, 127, 129, 130, 131, 132, 133
pelvic floor, vii, 1, 10, 11, 17, 18, 21, 28, 32, 33, 34, 110
peptides, 6, 7, 41, 45, 48
periosteum, 59
peripheral nervous system, 6
PET, 42, 50
pH, 40, 44, 64
pharmaceutical, vii, x, 1, 68, 127, 135, 136
pharmaceuticals, 10
pharmacokinetics, viii, 53
pharmacological treatment, 78
Philadelphia, 69, 100
phosphate, 44
photosensitivity, 60
physical exercise, 127, 129, 131
physical therapy, 18
physicians, 96
pigs, 38, 39
pilot study, 18, 23, 33, 109, 136, 140
placebo, 2, 3, 12, 13, 14, 18, 20, 22, 29, 30, 32, 34, 67, 71, 105, 107, 109, 110, 111, 140, 148
plasma membrane, 7, 40, 54
plastic surgeon, x, 2, 125, 126, 127
plasticity, 51, 117
platysma, 11, 115, 119
poison, vii, 2, 37, 38, 39, 40
polypeptide, 4, 136, 139
population, 4, 12, 20, 72, 114
positron, 42
positron emission tomography, 42
precipitation, 61
pregnancy, 27, 91, 137
preparation, 2, 3, 30, 49
prevention, 45
principles, 34
probe, 82, 93
professionals, 55
prognosis, 116
prolapse, 31
proliferation, 80
pronunciation, 114
prophylactic, 148
prophylaxis, 148
prostatitis, 107, 108
protection, 139
protein blocks, 47
protein design, 49
protein kinase C, 31, 44, 51
proteins, viii, 4, 5, 6, 8, 40, 54, 56, 57, 62, 68, 127, 139
pruritus, 60, 62
psychiatry, 133
ptosis, 60, 62, 63, 119, 145
public domain, 38
public health, 49
purification, viii, 54, 61

Q

quality of life, x, 12, 13, 14, 18, 24, 32, 44, 72, 76, 113, 116, 121

R

radiation, 83
radio, 74, 82
radiography, 92
radiotherapy, 74, 75, 76, 78, 83
radius, 68
rash, 27, 62
reactions, viii, 54, 60, 62, 63
receptors, 5, 6, 7, 8, 9, 30, 41, 44, 45, 74, 104, 139
reconstruction, 126, 132, 133
recovery, x, 43, 90, 113, 114, 116
recurrence, 146
recycling, 40, 41, 50
regeneration, 115, 120
regression, 76
rehabilitation, ix, x, 76, 113, 115, 116, 117, 122

relaxation, 26, 105, 147
remodelling, 89
renal failure, 12
researchers, viii, 13, 53, 55, 126
resection, ix, 87, 89, 92, 99
reserves, 16
residuals, 12, 19
resistance, 120
resolution, x, 18, 78, 80, 126
response, xi, 2, 3, 14, 20, 34, 43, 45, 47, 60, 68, 101, 106, 109, 111, 132, 136, 140, 143, 144, 146
responsiveness, 44, 59
rheumatoid arthritis, 137
rhinitis, 88
risk, viii, 15, 22, 23, 25, 26, 27, 28, 54, 65, 66, 78, 81, 111, 118, 132
risk factors, 111
room temperature, 62
root, 8, 45, 146
Russia, 62, 64

S

safety, vii, viii, 2, 3, 9, 20, 24, 25, 27, 40, 46, 54, 63, 66, 67, 68, 99, 118
salivary glands, 120
secretion, 38, 44, 74, 147
sediment, 4
seizure, 46, 137
self-esteem, 121
senescence, 66
sensation, 6, 63, 77, 104, 126, 147
sensitivity, 43, 44, 104
sensitization, 43, 44, 45
serum, 60, 65
serum albumin, 65
sex, 89
sexual activity, 18
shape, 41, 92
sheep, 38
side effects, ix, 12, 13, 16, 19, 20, 21, 23, 25, 26, 27, 28, 74, 87, 89, 90, 99, 107, 122, 136, 145
signaling pathway, 6
signs, ix, 3, 66, 77, 80, 90, 94, 100, 103
Singapore, 39
skeleton, 100
skin, 55, 60, 65, 66, 67, 74, 75, 78, 80, 81, 95, 146
smooth muscle, 6, 8
SNAP, 5, 6, 8, 39, 41, 43, 45, 57, 104
sodium, 58, 61, 63, 72, 80, 81
solution, x, xi, 56, 57, 58, 61, 64, 95, 118, 125, 128, 129, 131, 143, 145, 147
South Korea, 64
SPA, 55, 60
Spain, 103, 113
spastic, viii, 4, 10, 71
spasticity, 3, 10, 11, 17, 19, 29, 88
species, 4, 38
specifications, 141
sphincter, ix, 2, 10, 11, 12, 19, 20, 21, 25, 26, 29, 31, 33, 88, 103, 105, 108, 109
spinal cord, 2, 6, 7, 10, 12, 13, 26, 29, 30, 31, 32, 42, 45, 106, 109
spindle, 42, 43
spine, 84
splenius capitis, 76
splint, x, 94, 95, 100, 135, 136, 138, 139
spore, 4, 38
sprouting, 101
stability, 63, 139
stabilization, 116
stabilizers, 27
standardization, 32
states, 25, 43, 45
sterile, 56, 61, 63
stethoscope, 137
stimulant, 48
stimulation, 139, 146
stimulus, 120, 147
stock, 99
storage, 139
strabismus, vii, 2, 37, 39, 40, 48, 88, 126
stress, 108, 114
stretching, 104, 117
striatum, 42
stroke, vii, 37, 40
structure, 4, 29, 49, 68
subcutaneous injection, 81, 82
subcutaneous tissue, 128, 147
subjective experience, 92
substrates, 8, 41
supervision, 80
suprapubic, 15
surgical intervention, 17
surgical removal, 46
surgical resection, viii, 87, 89
survival, 8
susceptibility, 8, 84
swelling, 3, 63
Switzerland, 39
symmetry, x, 113, 116, 117, 118, 119

symptoms, ix, 1, 2, 3, 9, 12, 14, 15, 16, 17, 18, 19, 20, 21, 24, 27, 28, 32, 38, 46, 76, 100, 103, 105, 106, 107, 108, 109, 110, 111, 114, 138, 140
synapse, 6, 44, 54, 118
synaptic vesicles, 40, 41, 47
syndrome, ix, 6, 11, 12, 15, 16, 21, 23, 24, 25, 27, 30, 32, 33, 34, 45, 60, 88, 103, 104, 110, 111, 117, 122, 123
synthesis, 104, 139

T

Taiwan, 39
target, 8, 39, 52, 54, 93, 96, 118, 119, 140
teams, 80
techniques, 3, 19, 60, 81, 117, 144
technology, 48, 65
teeth, 95, 115
temperature, 10, 56, 59, 61, 63, 78
temporal lobe, 46
temporomandibular disorders, 140, 141
tension, 7, 18, 80, 115
terminals, 7, 8, 40, 43, 45, 96, 136, 139
testing, 64
tetanus, 4, 42, 50, 101
texture, 55, 60, 66, 67
theatre, 12
therapeutic benefits, 13
therapeutic effects, 9, 34
therapeutic use, 2, 29, 39, 50
therapeutics, 49, 52, 81
therapy, x, xi, 12, 25, 29, 33, 39, 43, 46, 51, 72, 74, 84, 85, 91, 94, 98, 99, 100, 101, 109, 110, 111, 118, 122, 125, 135, 136, 138, 139
tics, vii, 37, 40, 88
time use, 95
tissue, ix, 8, 31, 43, 44, 45, 54, 87, 93
tissue remodelling, ix, 87
tonic, 44
tooth, ix, 87, 90, 94
torticollis, 11
toxic substances, 39
toxicity, 4, 19, 29, 40, 48, 49, 136, 139
trade, 62
trafficking, 41, 42, 43, 48
training, 116, 117, 122
transection, 7
translocation, 4, 5, 40, 41, 47, 52
transmission, 2, 5, 6, 7, 39, 73, 96
transport, 42, 43, 45, 47, 50
trapezius, 76, 77
trauma, 78, 114
treatment methods, x, 125, 126, 135, 136
trial, 13, 17, 18, 20, 30, 32, 33, 34, 65, 106, 110, 111, 141
trigeminal nerve, 144, 145, 146
trigeminal neuralgia, vii, xi, 46, 81, 85, 143, 146, 147, 148
tumors, 78, 144
turnover, 54

U

UK, 1, 37, 39, 137
ultrasonography, 93
underlying mechanisms, 44
United States (US), 2, 56, 57, 61, 64, 66, 128, 137
urethra, 109
urinary bladder, 6, 8, 9, 30, 31
urinary retention, ix, 10, 11, 14, 15, 19, 26, 103
urinary tract, ix, 3, 6, 8, 15, 19, 27, 28, 32, 34, 103, 105, 107, 108, 109, 110, 111
urinary tract infection, 19
urine, 8, 14, 107
urothelium, 6, 8, 30

V

vaccine, 3
vagina, 18
validation, 66
variables, 13, 20, 25, 105
variations, 22
vascular cerebral stroke, vii, 37, 40
vascular system, 44
vasodilator, 7
vasovagal syncope, 63
velocity, 11
vertigo, 3
vesicle, 5, 6, 40, 41, 44, 45, 48, 49, 50, 54, 139
vesico-sphincter dyssynergia, ix, 103
vessels, 43
vision, 3, 62
voiding, 10, 12, 13, 14, 19, 20, 24, 26, 27, 28, 32, 33, 35, 105, 108, 109

W

weakness, 3, 26, 35, 60, 63, 107, 108, 141

World Health Organizaion (WHO), 3
Wisconsin, x, 113
wound infection, 4

Y

yield, 34

Z

zinc, 4, 6, 40, 54, 139
zygomatic arch, 72, 95, 147